I0138589

IMAGES
of America

WHEELER NATIONAL WILDLIFE REFUGE

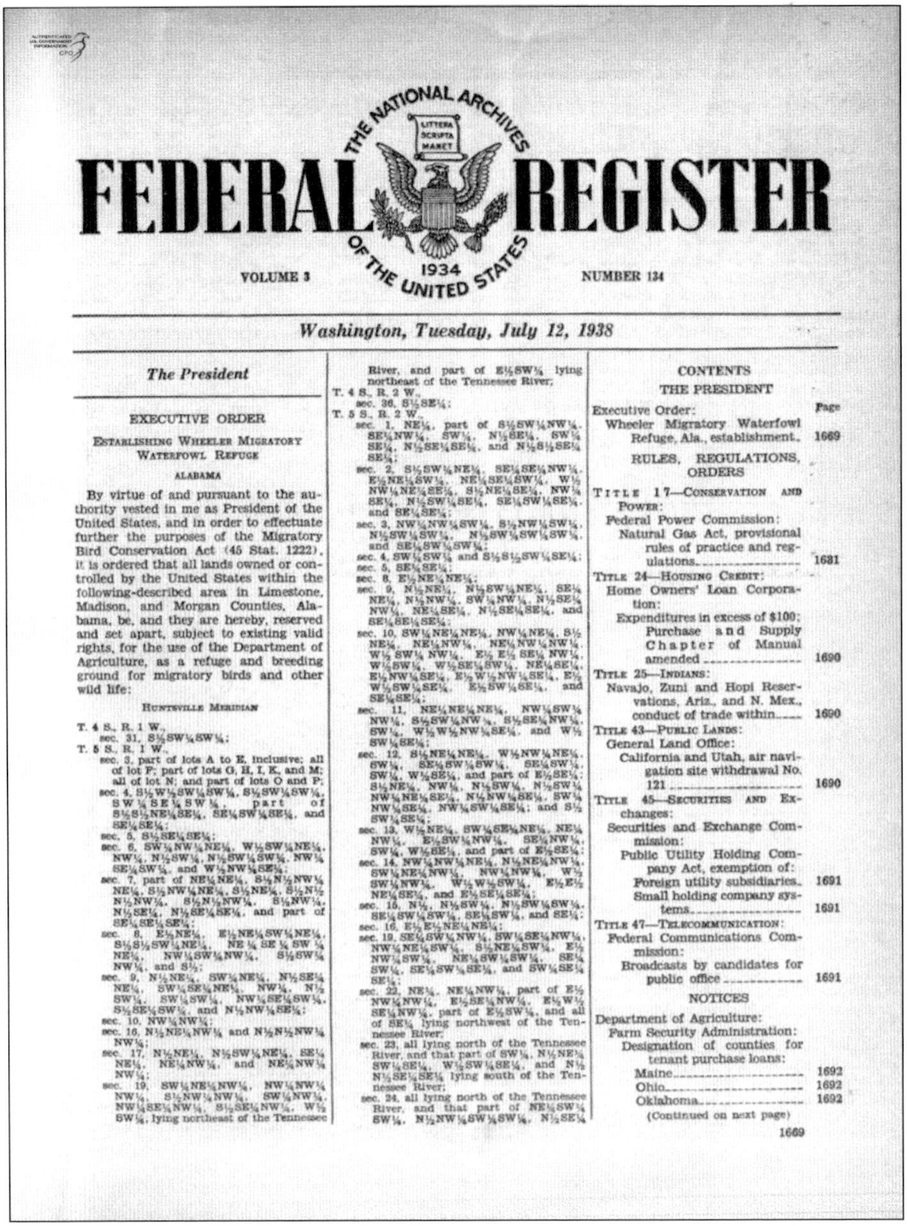

THE NATIONAL ARCHIVES OF THE UNITED STATES 1934 LITTERA SCRIPTA MANET

FEDERAL REGISTER

VOLUME 3 NUMBER 134

Washington, Tuesday, July 12, 1938

The President

EXECUTIVE ORDER

ESTABLISHING WHEELER MIGRATORY WATERFOWL REFUGE

ALABAMA

By virtue of and pursuant to the authority vested in me as President of the United States, and in order to effectuate further the purposes of the Migratory Bird Conservation Act (45 Stat. 1222), it is ordered that all lands owned or controlled by the United States within the following-described area in Limestone, Madison, and Morgan Counties, Alabama, be, and they are hereby, reserved and set apart, subject to existing valid rights, for the use of the Department of Agriculture, as a refuge and breeding ground for migratory birds and other wild life:

HUNTSVILLE MERIDIAN

T. 4 S., R. 1 W.,
sec. 31, S½SW¼SW¼;
T. 5 S., R. 1 W.,
sec. 3, part of lots A to E, inclusive; all of lot F; part of lots G, H, I, K, and M; all of lot N; and part of lots O and P;
sec. 4, S½W½SW¼SW¼, S½SW¼SW¼, SW¼SE¼SW¼, part of S½S½NE¼SE¼, SE¼SW¼SE¼, and SE¼SE¼;
sec. 5, S½SE¼SE¼;
sec. 6, SW¼NW¼NE¼, W½SW¼NE¼, NW¼, N½SW¼, N½SW¼SW¼, NW¼SE¼SW¼, and W½NW¼SE¼;
sec. 7, part of NE¼NE¼, S½N½NW¼NE¼, S½NW¼NE¼, S½NE¼, S½N½N½NW¼, S½N½NW¼, S½NW¼, N½SE¼, N½SE¼SE¼, and part of SE¼SE¼SE¼;
sec. 8, E½NE¼, E½NE¼SW¼NE¼, S½S½SW¼NE¼, NE¼SE¼SW¼NE¼, NW¼SW¼NW¼, S½SW¼NW¼, and S½;
sec. 9, N½NE¼, SW¼NE¼, N½SE¼NE¼, SW¼SE¼NE¼, NW¼, N½SW¼, SW¼SW¼, NW¼SE¼SW¼, S½SE¼SW¼, and N½NW¼SE¼;
sec. 10, NW¼NW¼;
sec. 16, N½NE¼NW¼ and N½N½NW¼NW¼;
sec. 17, N½NE¼, N½SW¼NE¼, SE¼NE¼, NE¼NW¼, and NE¼NW¼NW¼;
sec. 19, SW¼NE¼NW¼, NW¼NW¼NW¼, S½NW¼NW¼, SW¼NW¼, NW¼SE¼NW¼, S½SE¼NW¼, W½SW¼, lying northeast of the Tennessee River, and part of E½SW¼ lying northeast of the Tennessee River;
T. 4 S., R. 2 W.,
sec. 36, S½SE¼;
T. 5 S., R. 2 W.,
sec. 1, NE¼, part of S½SW¼NW¼, SE¼NW¼, SW¼, N½SE¼, SW¼SE¼, N½SE¼SE¼, and N½S½SE¼SE¼;
sec. 2, S½SW¼NE¼, SE¼SE¼NW¼, E½NE¼SW¼, NE¼SE¼SW¼, W½NW¼NE¼SE¼, S½NE¼SE¼, NW¼SE¼, N½SW¼SE¼, SE¼SW¼SE¼, and SE¼SE¼;
sec. 3, NW¼NW¼SW¼, S½NW¼SW¼, N½SW¼SW¼, N½SW¼SW¼SW¼, and SE¼SW¼SW¼;
sec. 4, SW¼SW¼ and S½S½SW¼SE¼;
sec. 5, SE¼SE¼;
sec. 8, E½NE¼NE¼;
sec. 9, N½NE¼, N½SW¼NE¼, SE¼NE¼, N½NW¼, SW¼NW¼, N½SE¼NW¼, NE¼SE¼, N½SE¼SE¼, and SE¼SE¼SE¼;
sec. 10, SW¼NE¼NE¼, NW¼NE¼, S½NE¼, NE¼NW¼, NE¼NW¼NW¼, W½SW¼NW¼, E½E½SE¼NW¼, W½SW¼, W½SE¼SW¼, NE¼SE¼, E½NW¼SE¼, E½W½NW¼SE¼, E½W½SW¼SE¼, E½SW¼SE¼, and SE¼SE¼;
sec. 11, NE¼NE¼NE¼, NW¼SW¼NW¼, S½SW¼NW¼, S½SE¼NW¼, SW¼, W½W½NW¼SE¼, and W½SW¼SE¼;
sec. 12, S½NE¼NE¼, W½NW¼NE¼, SW¼, SE¼SW¼SW¼, SE¼SW¼, SW¼, W½SE¼, and part of E½SE¼; S½NE¼, NW¼, N½SW¼, N½SW¼NW¼NE¼SE¼, N½NW¼SE¼, SW¼NW¼SE¼, NW¼SW¼SE¼; and S½SW¼SE¼;
sec. 13, W½NE¼, SW¼SE¼NE¼, NE¼NW¼, E½SW¼NW¼, SE¼NW¼, SW¼, W½SE¼, and part of E½SE¼;
sec. 14, NW¼NW¼NE¼, N½NE¼NW¼, SW¼NE¼NW¼, NW¼NW¼, W½SW¼NW¼, W½W½SW¼, E½E½NE¼SE¼, and E½SE¼SE¼;
sec. 15, N½, N½SW¼, N½SW¼SW¼, SE¼SW¼SW¼, SE¼SW¼, and SE¼;
sec. 16, E½E½NE¼NE¼;
sec. 19, SE¼SW¼NW¼, SW¼SE¼NW¼, NW¼NE¼SW¼, S½NE¼SW¼, E½NW¼SW¼, NE¼SW¼SW¼, SE¼SW¼, SE¼SW¼SE¼, and SW¼SE¼SE¼;
sec. 22, NE¼, NE¼NW¼, part of E½NW¼NW¼, E½SE¼NW¼, E½W½SE¼NW¼, part of E½SW¼, and all of SE¼ lying northwest of the Tennessee River;
sec. 23, all lying north of the Tennessee River, and that part of SW¼, N½NE¼SW¼SE¼, W½SW¼SE¼, and N½N½SE¼SE¼ lying south of the Tennessee River;
sec. 24, all lying north of the Tennessee River, and that part of NE¼SW¼SW¼, N½NW¼SW¼SW¼, N½SE¼

CONTENTS

THE PRESIDENT

(Continued on next page)

1669

Wheeler National Wildlife Refuge was created on July 7, 1938, when Pres. Franklin D. Roosevelt signed an executive order to establish the Wheeler Migratory Waterfowl Refuge to serve as "a refuge and breeding ground for migratory birds and other wildlife." The executive order was published in the July 12, 1938, *Federal Register*. Most of the order delineated the boundaries of the refuge; the executive order goes on for five pages. (Courtesy of Wheeler National Wildlife Refuge.)

ON THE COVER: The refuge entrance at the headquarters complex is pictured in 1955. Tree planting efforts are just starting to show results. This entrance along State Highway 67 now houses the modern headquarters building, a bunkhouse residence, and associated maintenance buildings. The area is now surrounded by a shady grove of mature hardwoods and pines. (Courtesy of Wheeler National Wildlife Refuge.)

IMAGES
of America

WHEELER NATIONAL
WILDLIFE REFUGE

Thomas V. Ress

ARCADIA
PUBLISHING

Copyright © 2019 by Thomas V. Ress
ISBN 978-1-4671-0432-6

Published by Arcadia Publishing
Charleston, South Carolina

Printed in the United States of America

Library of Congress Control Number: 2019947034

For all general information, please contact Arcadia Publishing:
Telephone 843-853-2070
Fax 843-853-0044
E-mail sales@arcadiapublishing.com
For customer service and orders:
Toll-Free 1-888-313-2665

Visit us on the Internet at www.arcadiapublishing.com

This book is dedicated to my wife, Roberta Thorn Ress, our daughter Sara Ress Wittenberg, and our son Michael Ress, and my late father, John V. Ress. My father's many hours spent hunting with me as a young boy in the fields of southern Indiana sparked my love of nature and the outdoors.

CONTENTS

ACKNOWLEDGMENTS

There are so many people to thank. Without their generous support, time, and encouragement, this book would never have been undertaken and completed.

Teresa Adams, supervisory park ranger, encouraged me to pursue this endeavor and provided access to the refuge's annual narrative reports and historical photographs. She dug through dusty files, old photographs, and overlooked storage areas to recover long-forgotten relics from the refuge's early days.

Ricky Ingram, Wheeler National Wildlife Refuge project leader, graciously allowed me virtually unfettered access to refuge archives and made sure my ramblings about refuge policies were correct.

Dwight Cooley, former Wheeler National Wildlife Refuge project leader, suggested that I undertake the writing of this book and provided a wealth of historical information about the refuge.

Susan Estes, Wheeler Wildlife Refuge Association board member, applied her keen teacher's eye in the review of the final manuscript.

Emery Hoyle, Richard Bays, and Daphne Moland gave additional historical input and checked my facts.

George Lee, Wheeler National Wildlife Refuge volunteer, provided some of his superb photographs of Wheeler's wildlife.

Finally, special thanks to my wife, Roberta, who puts up with my many hours exploring and enjoying the beauty of Wheeler National Wildlife Refuge.

Unless otherwise noted, all photographs in this book were obtained from and appear courtesy of Wheeler National Wildlife Refuge.

INTRODUCTION

If you could travel 80 years back in time and stand in the middle of what is now Wheeler National Wildlife Refuge (NWR), you would not recognize your surroundings. Back then, the lands of the current refuge in north Alabama's Tennessee Valley looked much different than today. Even by Depression-era standards, much of the valley was in sad shape in the 1930s. Many of the fields were worn out after decades of heavy farming, and much of the forested land had been extensively logged. Period photographs depict barren fields and eroded valleys and hillsides.

That started to change in 1934, when the Tennessee Valley Authority (TVA) began construction of a series of nine dams along the Tennessee River to provide flood control, year-round navigation, water supply, recreation, and hydroelectric power for the region. One of these dams was constructed between the towns of Decatur and Florence. TVA acquired land from local landowners to build Wheeler Dam and for the 67,100-acre reservoir behind it. Construction of the dam took three years, and the impounded waters behind the dam eventually formed Wheeler Lake.

Wheeler Lake changed the character of the area. Not only did the backwaters of the reservoir provide thousands of acres of fishing and other recreational opportunities, but Wheeler Dam also transformed the area in another dramatic way. The vast impounded lake with hidden back bays and sloughs was prime habitat for migrating ducks and geese on their way from the northern states and Canada to their wintering grounds in the South. Biologists realized the promise these changes held for wildlife, and within the Federal Bureau of Biological Survey (forerunner to the US Fish & Wildlife Service), proponents started pushing for the establishment of a wildlife refuge that would encompass part of Wheeler Lake and surrounding buffer land. Their efforts were successful, and on July 7, 1938, Pres. Franklin D. Roosevelt signed Executive Order 7926, establishing the Wheeler Migratory Waterfowl Refuge with a mission to serve as "a refuge and breeding ground for migratory birds and other wildlife."

The refuge was unique in two ways: it was Alabama's first national wildlife refuge and the first national wildlife refuge established on a multi-purpose reservoir. Up to that point, national wildlife refuges had been established in more traditional milieu: swamps, barrier islands, prairie wetlands, and wooded forests. The idea of establishing a refuge for wildlife in combination with an artificial lake seemed contradictory. Such a radical concept was considered folly by some, and the refuge was watched closely to see if it would succeed.

Any qualms were put to rest as the refuge proved to be an attraction for waterfowl. This was due in part to construction of backwater impoundments and the TVA's operation schedule, which involved dropping water levels in the spring and summer to control disease-spreading mosquito populations in surrounding communities. Lack of water in the impoundments during mosquito breeding season significantly reduced the mosquito population, and this periodic fluctuation had a beneficial side effect; the exposed shallows flourished with immense areas of natural grasses and were suitable for farming, producing food for wildlife. The abundance of food, along with the refuge's thousands of acres of sloughs, swamps, islands, and riparian habitat, soon drew large numbers of ducks, geese, and other waterfowl. The 1942 annual duck census cited a 50 percent increase in ducks over 1941.

Waterfowl may have taken to the area quickly, but the refuge did not have a grand debut, and 1938–1839 were years of making do with little resources. Eugene Cypert Jr. was temporarily transferred from White River Refuge in Arkansas as acting manager. Chester R. Markley soon took over as refuge manager and initially worked out of a rented office in the Old National Bank building in nearby Decatur before moving the headquarters into a Tennessee Valley Authority service building. A garage and storage building were leased from a local citizen to house equipment and supplies. With a meager budget, initial efforts involved scrounging equipment and supplies to run the refuge. Management borrowed a Chevrolet sedan from White River Refuge and picked up a surplus Plymouth sedan from the Bureau of Internal Revenue. Management even had to cadge metal boundary signs from Sabine National Wildlife Refuge in Louisiana to mark the new refuge boundaries.

From this shaky start, herculean efforts were undertaken to transform the refuge into a natural oasis that would continue to attract wildlife. One of the first intensive efforts on the newly established refuge was planting seedling trees and reseeding eroded areas; almost 500,000 seedlings were planted over the next three years. Since the refuge staff consisted of only six people, much of this work was performed by Works Progress Administration (WPA) workers and Civilian Conservation Corps (CCC) crews from a nearby CCC camp in the town of Hartselle. In 1940, CCC personnel embarked on the construction of the first infrastructure for the refuge, breaking ground on a headquarters building, a service building, and a manager's residence on Flint Creek Island just outside of Decatur, and another staff residence house on a satellite compound near the town of Triana. Flood control structures, bridges, and roads followed in quick succession. A second residence was later constructed at the refuge headquarters.

With this careful shepherding, the lands transformed in a relatively short time. In the former farmlands bordering the lakes, young saplings took hold, and wildflowers and native plants sprouted. The depleted fields and valleys slowly reverted to their natural state. As wildlife and woods reappeared, so too did people. The refuge promptly became a popular recreation spot, and its location in a populated semi-urban area near Decatur, Huntsville, and Muscle Shoals attracted eager visitors. Refuge personnel reported about 1,000 visitors per day during the summer of 1940 and 100 per day through the rest of that year.

Visitor numbers decreased for the next four years during World War II when many local residents went overseas to fight or worked overtime hours in factories supporting the war. Historical records document that drop in visitation. The refuge's 1942 annual report states that visitors decreased 80 percent because "the use of motorboats for pleasure has practically ceased due to gasoline rationing" but visitors still came and the reports state that "even during the stress of wartime living . . . the public still finds time . . . to do some fishing and get out of doors." Also during this period, the CCC camp was shut down and the refuge budget was cut, halting infrastructure development. The war had another effect on the refuge; in 1941, the US Army established Redstone Arsenal in nearby Huntsville, and 4,085 acres of the refuge's total 35,000 acres were incorporated into Redstone's boundaries.

But the war did not affect the birds, who continued to find the new refuge to their liking. Waterfowl populations soared. In 1943, refuge personnel recorded 28,000 to 33,000 migrant and wintering birds and noted in the refuge annual report that this was "a 100% increase over last year." Any doubts about the efficacy of locating a wildlife refuge on a multi-purpose reservoir disappeared. Waterfowl populations continued to swell over the next decades, so that by the 1970s, the refuge was home to the nation's southernmost, and Alabama's only, significant concentration of wintering Canada geese and the state's largest duck population.

The increase in waterfowl resulted in a related increase in visitors, and by 1980, a modern visitor center was built just down the road from the original refuge headquarters compound. The Lawrence S. Givens Visitor Center (named after an early refuge manager) houses static historical and wildlife displays and has a large auditorium and classroom. The two-story observation building constructed in 1973 overlooks a large display pool that abounds with thousands of ducks, geese, and other waterfowl during the peak of the winter migration season.

Wheeler National Wildlife Refuge has become an important component in maintaining viable waterfowl populations along the Mississippi Flyway, and in the past, as many as 60,000 geese and 125,000 ducks have spent the winter there. In the 1990s, the winter goose populations declined significantly due to a number of environmental factors, and now fewer than 1,000 Canada geese show up annually. Similarly, duck populations have declined and now number about 60,000–75,000. These declining goose and duck numbers are countered by an increase in numbers of majestic sandhill cranes that now overwinter on the refuge. The refuge remains a vital overwintering destination for many species of waterfowl and is one of the premier birding destinations in Alabama.

The refuge is now one of the major outdoor recreational attractions in Alabama, drawing 600,000 visitors annually to view its teeming waterfowl populations and enjoy its natural beauty. Recreational opportunities include fishing, kayaking, hunting, photography, hiking, biking, and birdwatching. In recent years, the refuge has hosted large numbers of migratory waterfowl including over 20,000 sandhill cranes, 75,000 ducks, 4,000 geese, and as many as two dozen endangered whooping cranes.

Much more has changed since the establishment of the refuge in 1938. Today, Wheeler NWR is located in a rapidly growing urban area and has become a natural oasis among ever-expanding development. The refuge is surrounded by the hustle and bustle of modern life—shopping centers, highways, gas stations, neighborhoods, and restaurants. But within the boundaries of the refuge, time has slowed. Deep within its 35,000 acres, none of this intrudes. Sleepy southern swamps, lofty cypress and southern pine trees, vast spreads of blooming lotus, and misty sloughs halt the crush and rush of civilization. Rafts of hundreds of ducks and geese and raucous flocks of sandhill cranes crowd the fields and sloughs. Coyotes, bobcats, deer, bald eagles, hawks, beavers, river otters, muskrats, alligators, and a myriad of other animals thrive on the refuge.

Wheeler NWR is a valuable remnant of wild Alabama that is cherished by outdoor lovers and wildlife enthusiasts. A walk in the refuge offers invigorating respite from the hubbub of the surrounding towns and housing areas.

One

THE EARLY YEARS

A Civilian Conservation Corps bulldozer moves dirt at the refuge headquarters site in preparation for construction. Almost all of the construction, land clearing, and tree planting from the refuge's creation in 1938 until 1942 (when the CCC was disbanded due to World War II) was performed by CCC or Works Progress Administration crews.

This 1940 photograph shows a badly eroded bank on the refuge, a typical sight at that time. The land had been heavily farmed and was depleted. Deep erosion areas like this were common. Prior to construction of dams on the Tennessee River, much of the land along the river had been heavily farmed and depleted. By 1940, deep erosion areas were common, and much of the natural river bank was eroded.

More erosion in the Triana area of the refuge is apparent in this 1941 photograph; severe erosion and barren ground were the norm. Even though there was a near-total lack of trees and ground cover in 1941, this area is now heavily forested.

These stumps are all that remain after the TVA harvested timber from the future pool area of Wheeler Lake. All trees, structures, and other obstructions were removed before the pool was filled to ensure safe passage for boats. Note the enormous size of the stumps.

Pine seedlings were planted by hand and these workers, probably a CCC crew, are hard at work planting in 1940. This was during the early stages of a multi-decade effort that planted thousands of seedlings and changed the character of the landscape. Between 1933 and 1942, there were about 30 CCC camps operating in Alabama employing about 66,000 men.

A mule-drawn rig is mowing a wetland area in 1941. Mowing was performed in the early days of the refuge for mosquito control. The CCC crews were primarily employed in construction projects while WPA crews were mostly used for agricultural projects, so this rig was probably operated by WPA workers.

This is a wooden bridge on Truck Trail constructed by CCC workers in 1941. This area of the refuge was primarily agricultural and was crisscrossed by creeks and ditches. Bridges were built to provide staff and visitors access to remote areas of the refuge.

This tractor is preparing the site of the boathouse on Flint Creek in 1940. Site clearing and construction were performed by CCC crews. The boathouse was built to house watercraft used by refuge personnel. It was demolished in 1978.

Civilian Conservation Corps crews built this bridge in 1940 on Rockhouse Landing Road. The CCC crews were housed in three camps located in Hartselle, Athens, and Huntsville. In 1940, these crews built 12 miles of graded truck trail, cleared another 16 miles of road, and constructed two permanent bridges and six temporary patrol bridges.

When the refuge first opened, managers borrowed vehicles from other refuges or obtained surplus vehicles from other government agencies. The Department of Interior logo on the door identifies this as one of the first trucks obtained by the refuge. Almost all of the vehicles obtained by the refuge in the early years were military surplus.

The manager's residence at refuge headquarters is ready for occupancy shortly after completion by CCC crews in 1941. This building is still standing and is used for storage. The CCC worked on two types of projects for Wheeler: improvement of wildlife habitat and construction of facilities. Their projects included construction of buildings, dams, dikes, and other water impoundment structures, planting of cover vegetation and trees, and stabilization of streambanks.

The wood frame of the refuge office building was photographed shortly after construction started. This building was one of the first erected on the refuge. Construction started in 1940 and was completed in 1941.

The office building is under construction in 1941. This was one of about six buildings including a manager's residence, employee residence, service building, and equipment building in the headquarters complex that were all completed by CCC crews in 1940–1941. These were the first buildings constructed on the new refuge. This building served as the headquarters office and is shown completed in 1941 below. It was demolished and replaced by a new office building in 1978.

Numerous bridges were constructed to provide entry to areas that were inaccessible due to creeks and eroded ditches. This bridge was built across Black Branch in 1940. The bridge, built with steel beams and rock supports, replaced the temporary wooden log bridge in the foreground. This photograph was taken in 1940, documenting this as one of the first bridges built on the refuge. The rock supports are still in use today.

A Dodge Power Wagon parks on the newly rebuilt Ginhouse Branch bridge in 1949. This bridge replaced one that was originally constructed by CCC crews earlier in the decade.

A worker, probably from the WPA, is plowing a firebreak by mule team before burning operations in 1940. Controlled burns were one tool used by refuge personnel to stimulate natural plant growth. Great effort was put into habitat improvement in these early years. In 1940 alone, CCC crews completed 32 acres of quail habitat improvement areas, planted 500,000 trees and shrubs, constructed 1,000 check dams for erosion control, and planted 140 experimental food plots for waterfowl.

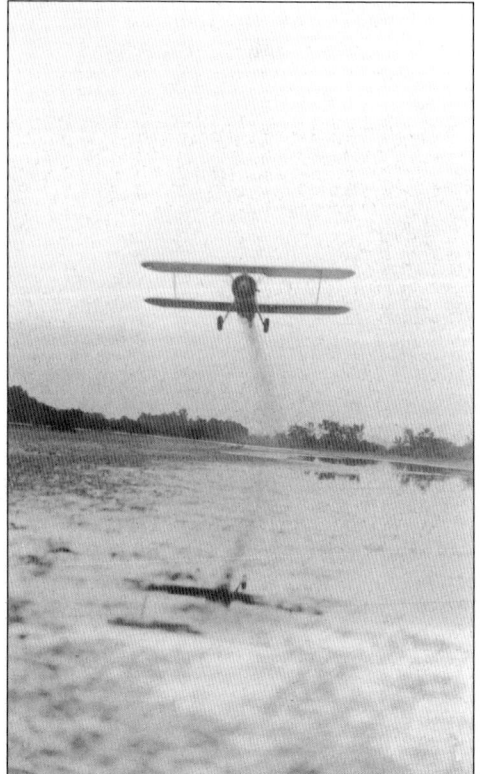

Ryegrass was sown by airplane on barren fields and mudflats. Airborne sowing was successful but expensive, and was used primarily on mudflats like this that were not easily sowed from the ground. This photograph is from around 1947.

An old canoe, believed to be of Native American origin, was discovered in the Tennessee River near the Triana headquarters in 1947. The current whereabouts of this artifact is unknown.

Refuge boats that were used for law enforcement patrols and wildlife management were housed in this boathouse, built in 1940. The boathouse was constructed at the headquarters complex and towed down Flint Creek to the Tennessee River and then upriver to the Triana station.

The *Limpkin,* part of the refuge "fleet" of two boats, was used by the refuge for law enforcement patrolling and waterfowl management. The speedboat below was used for law enforcement patrols. Enforcement of hunting and fishing regulations was a significant part of staff duties in the first years after the refuge was created.

Commercial mussel farming was common in the Tennessee River in the 1940s and was a major economic activity. In 1948, over 700 tons of mussels were harvested from refuge waters alone. This fleet of mussel digging boats is at work on the river in 1946. Mussel farming continued into the mid-1960s. In later years, as mussel populations plummeted, large-scale farming was replaced with smaller mussel diving operations.

A mussel fisherman is removing mussels from hooks. The apparatus is called a braille and is dragged across the river bottom over mussel beds. Feeding mussels clamp down on the hooks and hold tight until pulled off by the operator.

This is called a "cook out site." Mussel fishermen were permitted cook out sites on the refuge at Duncan Hill and Mussel Camp. Mussels were cooked, and the meat was separated from the shells. The meat was sold for human and animal consumption, and the shells were used for making buttons.

A 40-ton pile of mussel shells is ready for shipment to a button factory in 1946. Shells sold for $40 per ton.

Panning for pearls was a sideline activity to mussel farming. This father and son team are washing and screening soil from under a mussel shell–cleaning rig to recover slugs, or misshapen pearls, lost during the cleaning process.

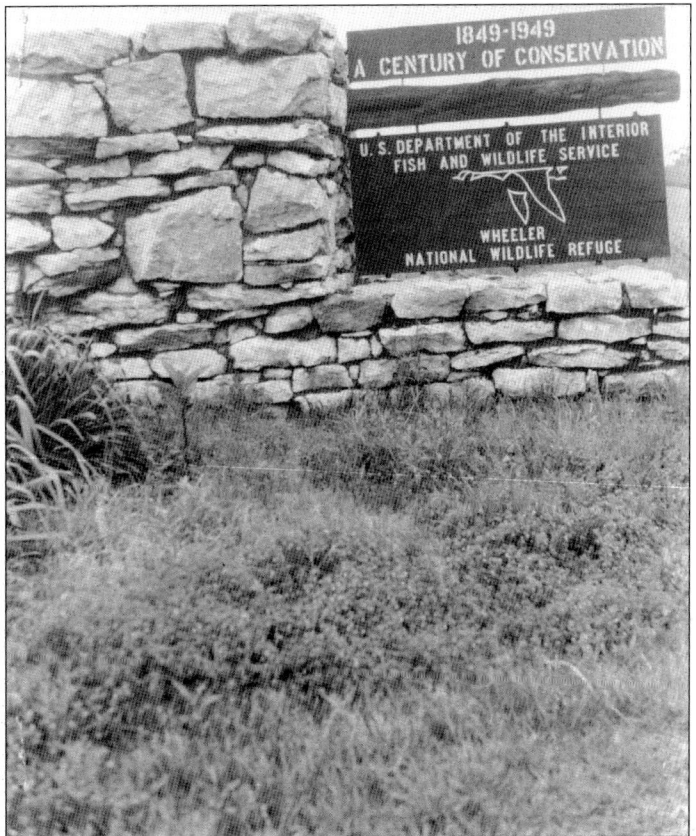

1849-1949
A CENTURY OF CONSERVATION

U. S. DEPARTMENT OF THE INTERIOR
FISH AND WILDLIFE SERVICE

WHEELER
NATIONAL WILDLIFE REFUGE

This "A Century of Conservation" sign was erected at the entrance to refuge headquarters on Highway 67 in 1949 to commemorate the founding of the Department of Interior in 1849. Highway 67 was a two-lane road at that time.

Patrolman Henry Grammer is pictured at the base of a massive winged elm in 1940 that somehow escaped harvesting. The survival of this tree was an exception; practically all harvestable timber had been cleared from the area before the refuge was established in 1938.

A local farmer gets in an early wheat harvest on Flint Creek Island in 1941. Cooperative farming agreements have long been an integral part of management on the refuge and continue to this day.

This is a c. 1940s photograph of the refuge entrance at the headquarters complex. Note the barren fields, reflective of the infancy of the tree planting efforts.

Wheeler Dam was constructed by the Tennessee Valley Authority and took three years to complete. It is one of nine dams on the river owned and operated by the TVA and was built as part of a New Deal–era initiative to improve navigation on the river and bring flood control and economic development to the region. Construction began on November 21, 1933. The project required the purchase of 103,400 acres of land and the removal of 840 families, 176 graves, and 30 miles of roads. At its peak, the project employed 4,700 workers and was completed on November 9, 1936, at a cost of over $87 million. Wheeler Lake, formed by the Tennessee River behind the dam, constitutes 16,000 acres of Wheeler National Wildlife Refuge. The dam is named for Civil War general and US congressman Joseph Wheeler. (Photograph by Dwight Cooley.)

A CCC construction crew is building a boat launch ramp on Flint Creek. This was part of the initial infrastructure construction that took place in 1940–1941. The majority of the construction projects on the refuge at this time were concentrated around the headquarters site on Flint Creek near State Highway 67. In addition to the obvious construction that was going on, crews built a well house, water system, and sewer system. A secondary headquarters complex was also constructed at Triana, Alabama.

This dragline is working in the Rockhouse Road area, excavating soil in 1941 for construction of a water control structure for a dewatering unit at Buckeye Pond, a WPA project. Dewatering units were areas of the refuge that were controlled by pumps and water-control structures to allow for management of water levels to enhance waterfowl habitat.

This crew (probably a CCC construction crew) is pouring concrete for a project in the Flint Creek area. This was part of a water impoundment facility. The photograph was taken in 1940.

A TVA employee clears thick vegetation from waterways using a Hockney aquatic mower, a floating machine used to cut vegetation. Lotus plants choked many of the backwaters of the Tennessee River in the refuge. This machine cut the plants and opened up the waterways for boating.

The headquarters office building is pictured after a rare southern snowfall in 1958. The results of the tree planting program are evident in the tall trees in front of the building. The structure to the right of the office is the equipment building, and the boathouse is visible on the far right in the background.

Two

WILDLIFE

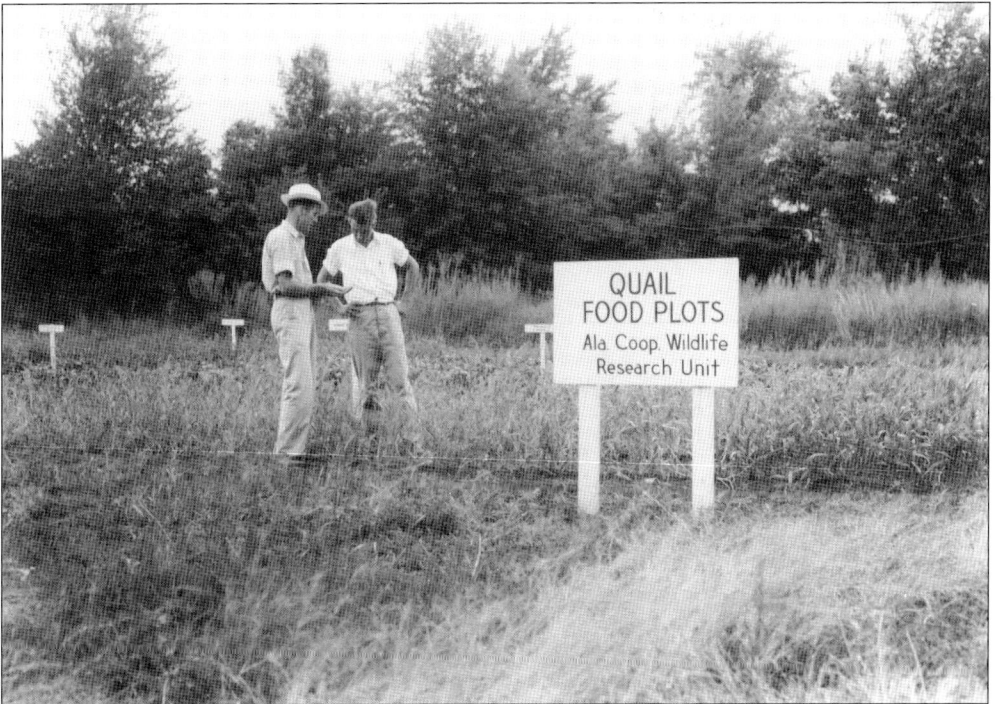

Dr. Alan Pearson (left) and James Dukes of the Alabama Cooperative Wildlife Research Station are inspecting an experimental wildlife food plot. Early efforts concentrated on increasing wildlife habitat areas on the marginal lands on the refuge. Numerous experimental plantings were made in the White Springs, Bradford Sinks, and Flint Creek areas.

Superintendent F.E. Cox of the Elk River Fish Hatchery is planting fingerlings in refuge waters in 1941. Game fish were stocked on the refuge to enhance fish populations and to encourage recreational fishing.

Refuge personnel, from left to right, T.P. Sandlin, G.C. Bishop, and E.N. Waldrep remove Canada geese from a trap. These geese were banded and released. A total of 1,655 ducks and geese were banded during the 1962–1963 trapping season. Note the large number of waterfowl flying over the horizon in the distance. These men are working at the display pool before the observation building was constructed.

In November 1962, the Alabama Conservation Department released 91 Iranian pheasants in the White Springs Dike area. Here, state biologist James Keeler prepares for release. This release was part of a state program to introduce the non-native bird to north Alabama as an additional quarry for bird hunters. Birds were also released in the nearby Swan Creek Wildlife Management Area, a state-owned hunting area.

Refuge patrolman Henry Grammer releases an Iranian pheasant in the White Springs area.

33

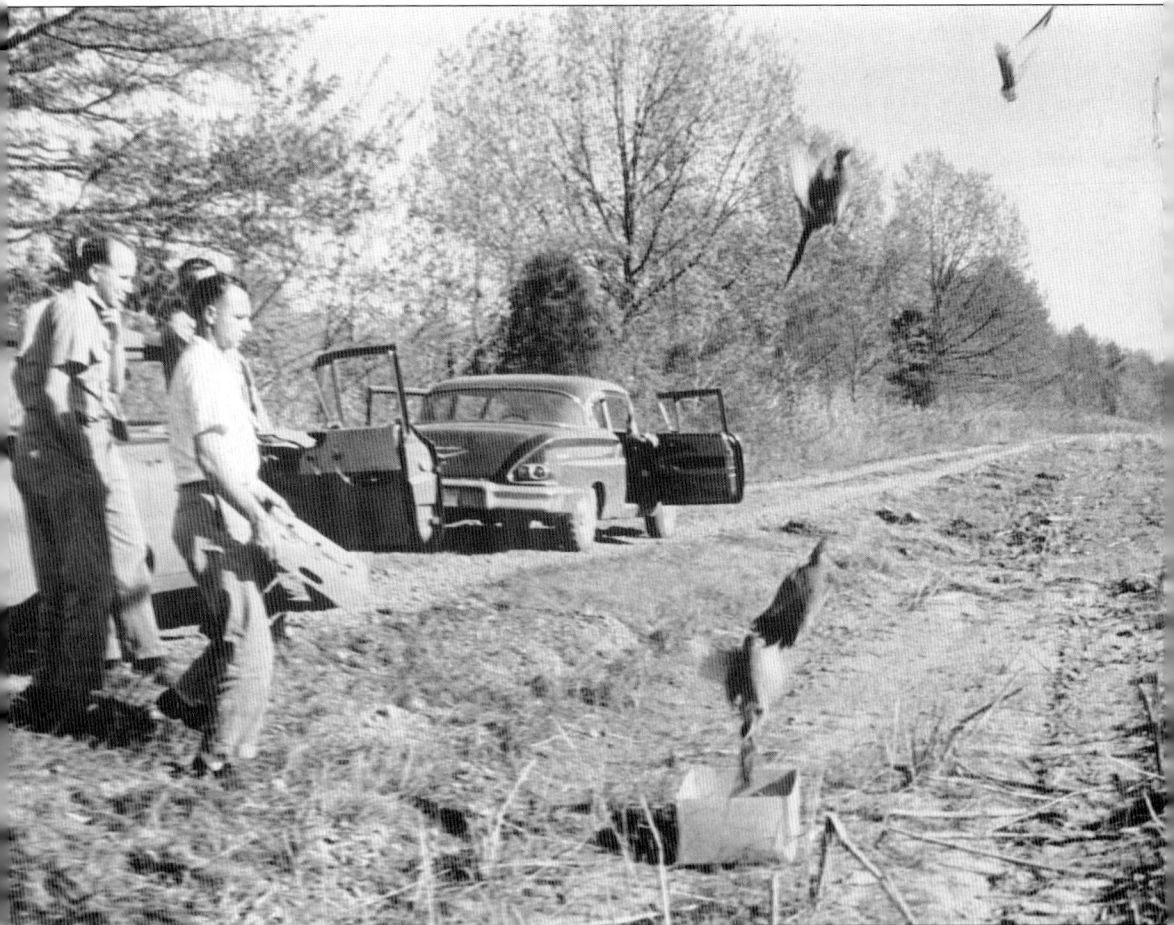

Efforts continued to introduce Iranian pheasants in 1963 with the release of 70 birds. An Alabama Conservation Department biologist releases the birds while refuge clerk Harvie Fowler, on the left, and state biologist Jim Keeler watch them fly. The pheasants persisted for a few years, but the population eventually died out.

Wheeler's duck and geese populations peaked in the mid-1960s, when scenes like this were common. Duck populations reached 125,000, and in 1963, there were 60,000 Canada geese reported on the refuge. Numbers of Canada geese have declined precipitously since. Refuge employee Richard Bays described the skies as "black with birds." Peak waterfowl numbers of over 120,000 were recorded in 1996.

Wood duck boxes were constructed by a Huntsville sportsmen's club and donated to the refuge in 1966. From left to right, refuge personnel Henry Grammer, unidentified, Emmett Waldrep, and Tom Sandlin erected the boxes. Wood ducks are often spotted on the refuge in densely wooded areas. Refuge staff continue to maintain over 100 wood duck nesting boxes on the refuge.

This huge beaver lodge discovered in the Beaverdam Creek Swamp in 1968 was 18 feet in diameter and seven feet in height. Beavers are a common sight on the refuge, but in the first few years after refuge establishment, they were scarce. Beavers were trapped in other locations by state biologists.

Patrolman Henry Grammer holds a dead female mallard found on the refuge in 1940. As the waterfowl population increased, natural mortality numbers followed. The winter of 1940 was very cold, as evidenced by the snow cover in this picture, and probably contributed to increased mortality that year.

This is a typical winter sight on the refuge. Wheeler is located on the eastern edge of the Mississippi Flyway, an important migration corridor for waterfowl. Migrating waterfowl follow this flyway from Canada and the northern tier of states down the Mississippi River Valley as far as the Gulf of Mexico. Many birds stop at Wheeler NWR and spend the winter, and thousands of geese, ducks, and cranes can be observed congregating and feeding in the fields and wetlands. Migrating birds begin showing up at Wheeler in October and stay until early to late March.

Refuge personnel have banded and studied various species of waterfowl over the years. This Canada goose is collared (a form of banding) during a data collection effort that took place in 1988. Canada geese are no longer banded on the refuge, but a small number of wood ducks are still trapped and banded annually.

In 1983, these ospreys were hacked by refuge employees in cooperation with TVA and state conservation personnel. Hacking is the process of introducing new fledgling birds to an area where they have been missing or rare. The birds were hacked on a platform at the display pool by the observation building. There is still an osprey platform in this location. Once rare, ospreys are now common on Wheeler, and the refuge has at least a half dozen active nests. Ospreys prey primarily on fish, and prior to the construction of Wheeler Dam, their numbers were low. The vast water of Wheeler Lake means more fish, which attracts more ospreys.

This picture of an osprey nest on a Tennessee Valley Authority transmission tower was taken in 1989. This nest, or a replacement in the exact location, has been in more or less constant use since at least that date and is still an active nest today.

Biological service technician Faye Blankenship and forester Richard Bays band ring-necked ducks from a deep water trap in White Springs. Banding efforts have varied over the years. In 1983, when this photograph was taken, refuge staff banded 23 Canada geese, 939 mallards, 79 ring-necked ducks, 109 wood ducks, 16 redheads, 11 canvasbacks, 189 black ducks, and 296 American wigeons. Banding operations on the refuge are now limited almost exclusively to wood ducks.

Visitors are often surprised to discover that there are alligators on Wheeler NWR. Historical documents record sightings of alligators in the Tennessee Valley for over a century. Alligators were spotted in the Tennessee River as early as the 1890s, and they have been infrequently sighted ever since. There is a sustaining population on the refuge, and sightings are not uncommon. This four-footer was stranded during remedial habitat work in Huntsville Spring Branch in 1987 and was captured, relocated, and released in Blackwell Swamp.

In 1979, federal biologists made a decision that still sparks discussion and misinformation to this day. In June of that year, 55 alligators were trapped at Lacassine NWR in southern Louisiana and transplanted to Wheeler NWR. A common belief among area residents is that the gators were brought in to control excessive beaver populations on the refuge, but the relocation was part of a larger effort by the US Fish & Wildlife Service (USF&WS) to ensure that scattered populations of what was at that time an endangered species would increase its chances for recovery should some calamity strike core populations farther south. From this action, many locals believe that federal officials are solely responsible for the presence of alligators in north Alabama. This photograph shows a federal employee releasing one of the 55 alligators.

The release of the Lacassine NWR alligators was generally well-received by the public, and local media coverage was favorable. But a few citizens were concerned, and a local congressman took exception to the release and demanded that the alligators be removed. Given the vastness of the 35,000-acre refuge and the tendency of gators to disperse to the most isolated and remote areas, this was a quixotic demand. However, the refuge undertook an earnest effort to recapture the transplanted animals, and in June and July 1980, teams of experienced wildlife personnel from Wheeler, other refuges, and USF&WS offices worked for 10 consecutive nights to attempt the recapture. The effort was futile; only two animals were caught and returned to Louisiana. George Chandler, who was refuge manager at Mississippi Sandhill Crane NWR and Tensas River NWR, poses with one of the recaptured alligators.

Requests continued to come in for personnel to capture alligators. Refuge equipment operator David McCathren (left) and an unknown employee hold a three-footer that was caught in 1991 in eastern Morgan County and released in Blackwell Swamp. Also that year, a five-and-a-half-foot-long alligator was caught in the Massey community, and a seven-footer was caught near Florette. Both were released on the refuge. Off-refuge alligator reports are now handled by the Alabama Division of Wildlife and Freshwater Fisheries.

Both gray and red foxes are common on the refuge. This is a particularly handsome gray fox. Coyotes are also commonly seen on the refuge, and bobcats are occasionally spotted.

Two red-shouldered hawk fledglings fell out of their nest during a storm and were rescued by refuge personnel. Various species of raptors are known to live on the refuge, including ospreys, bald eagles, red-tailed hawks, red-shouldered hawks, American kestrels, and numerous owl species.

Canada geese are silhouetted moving across a hunter's moon in 1982. Canada geese numbers have decreased dramatically since the 1980s, but greater white-fronted geese numbers have increased. The refuge also enjoys a sizable population of overwintering snow geese.

Populations of spring pygmy sunfish (top), a threatened species, and Tuscumbia darters (bottom), a vulnerable species, have been confirmed in refuge waters. The known range of both of these fish is limited to a few locations in the tributaries of the Tennessee River. The refuge is home to 13 endangered or threatened species.

Auburn University graduate student John Thompson releases a wood duck that has been tagged with telemetry equipment as part of a research project in 1984. In an effort to determine whether locally hatched birds remained in the vicinity or migrated, 64 ducks were tagged.

From left to right, Carolyn Garrett, Richard Bays, and Anita Bowman hold a pair of ruddy shelducks, a species native to Europe and Asia, that were observed on the refuge in 1982 and caused a stir with local bird enthusiasts. In 1983, these two ducks were captured, banded, and released by refuge staff. They were never seen again.

A Eurasian wigeon, a rare bird in Alabama, and the first recorded sighting at the refuge, caused a stir among birdwatchers in 1982. Although still an uncommon sighting on the refuge, a lone individual or two have made sporadic appearances in recent years, and they still cause excitement for birdwatchers.

Hawks, bald eagles, and owls are common refuge residents. This young barred owl was brought to the visitor center by a local resident and was sent to the Alabama Wildlife Center in Birmingham for care and rehabilitation.

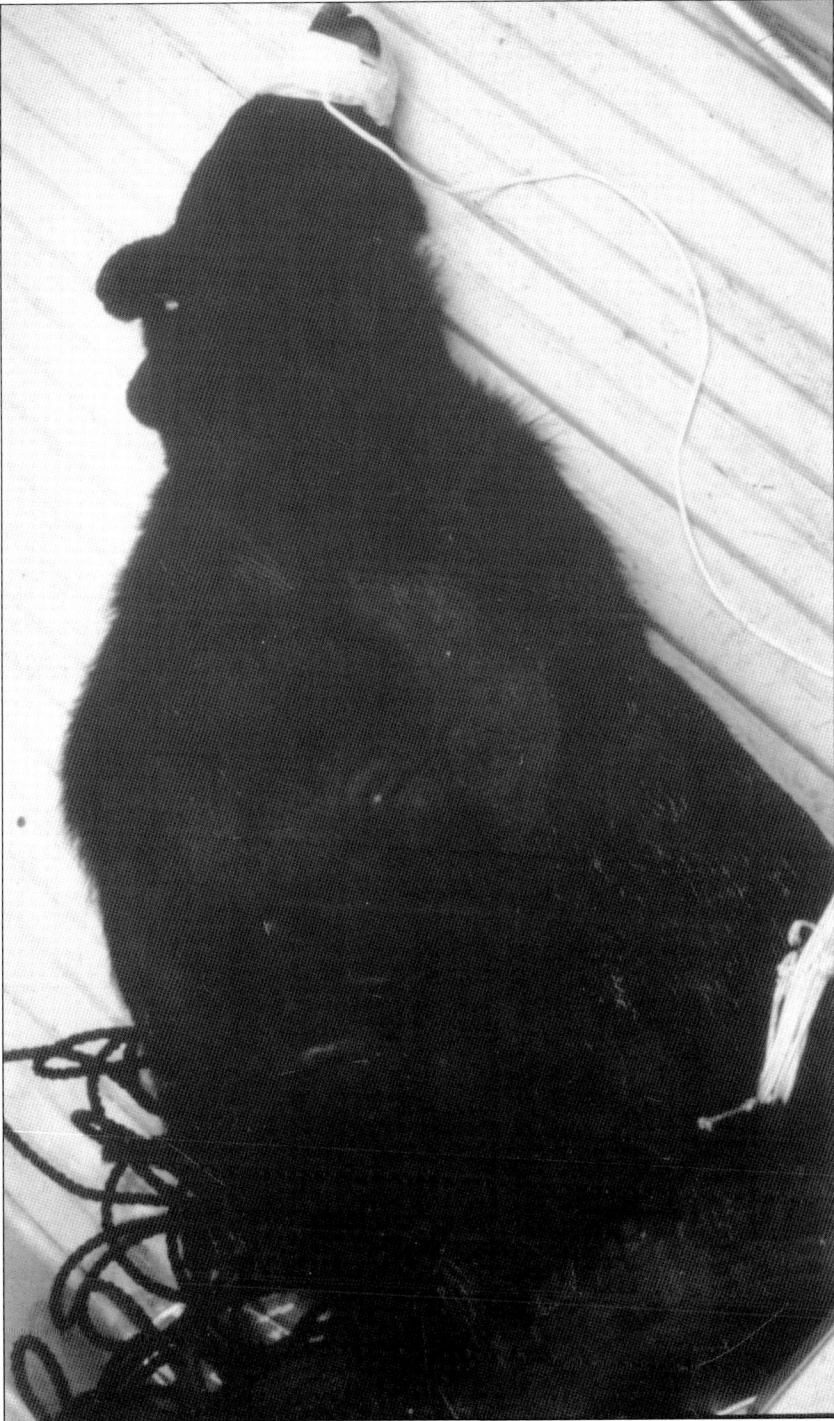

A black bear showed up in the backyard of a local resident in 1985. Refuge personnel assisted state conservation officers in tranquilizing it and transporting it to the Birmingham Zoo. It was later released in the southern part of Alabama. There are no resident bears on the refuge, but an occasional transient animal may pass through.

Beaver dams were occasionally destroyed to protect roads and facilitate drainage in dewatered units. This photograph shows US Fish & Wildlife Service employees Cindy Dohner and Faye Blankenship preparing dynamite charges to blow up a dam. Beaver dams can cause flooding over roads and onto neighboring farmlands.

There she blows! This beaver dam was blown up to resolve a flooding problem in the Beaverdam Peninsula area. Beaver dams are not destroyed unless they cause flooding issues on the refuge or neighboring farmlands. Destruction using dynamite is the easiest and quickest way to remove the dams.

In contrast to Canada geese populations, which have been decreasing for decades, Sandhill crane numbers are increasing at Wheeler NWR. In the early 1980s, a few sandhill cranes were spotted on the refuge. This was a rare sighting at the time since there had been no cranes in the Tennessee Valley for decades. Prior to 1997, Sandhill cranes occurred only rarely and in very small numbers on the refuge. In 1997, a total of 26 were observed, and by 2002, the number wintering on the refuge had increased to almost 400. During the winter of 2017–2018, the number peaked at 20,000. The cranes begin arriving in late November, and numbers peak in mid-to-late January. (Photograph by George Lee.)

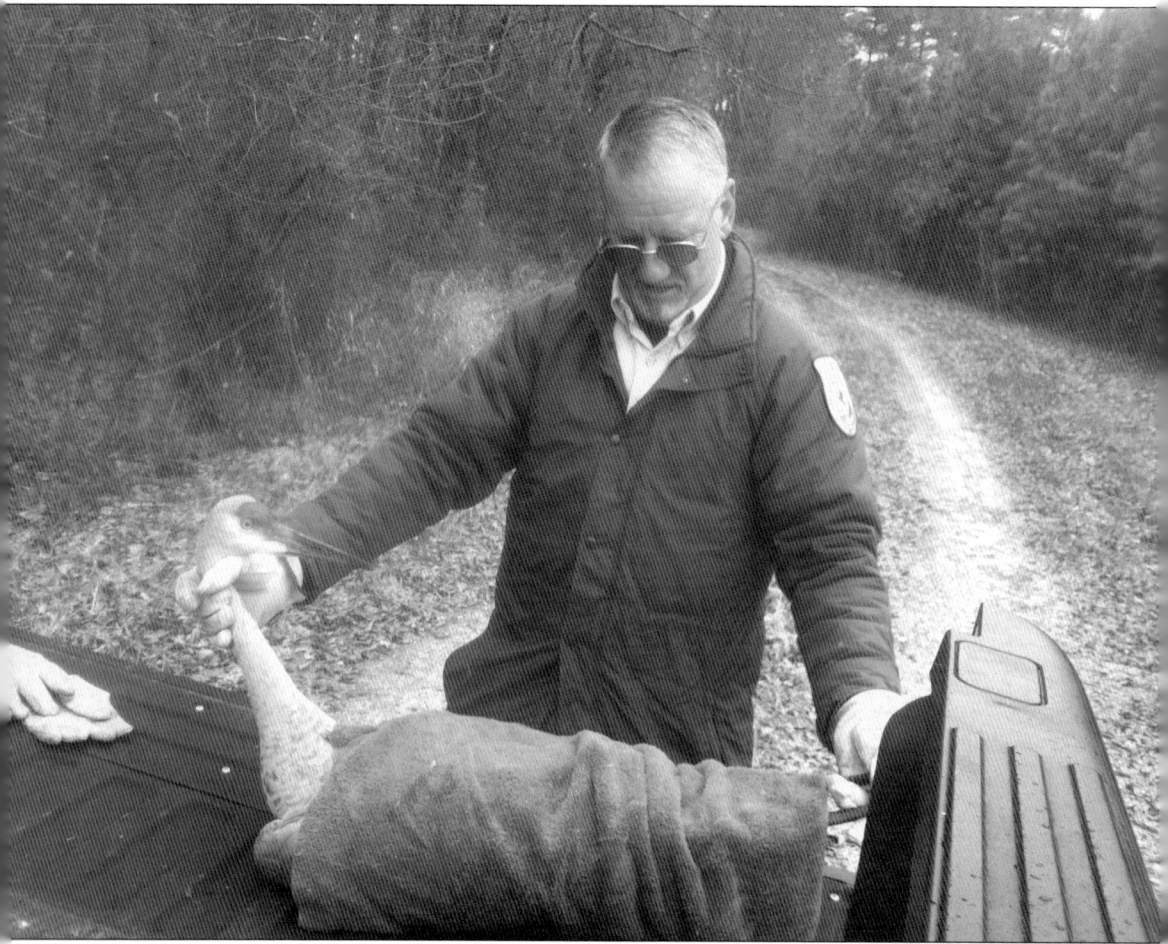

Ranger Kevin Hamrick assesses an injured sandhill crane that was captured in the Dinsmore Slough area. This bird had been observed unable to fly, and its survival was doubtful. In the vast majority of cases, nature is allowed to run its course, but staff feared this bird was sick, and it was taken to determine the cause. (Photograph by the author.)

In January 2005, two rare whooping cranes from one of the Operation Migration cohorts were discovered on the refuge. These were the first whooping cranes recorded on Wheeler. Since 2007, whooping cranes have wintered each year on the refuge. During the winter of 2017–2018, up to 27 whooping cranes spent a portion of their winter on the refuge. In recent years, they have frequented areas around the visitor center and wildlife observation building, allowing thousands of refuge visitors the opportunity to view them. Whooping cranes are an endangered species and one of the rarest birds in North America, with a total population of less than 850. (Photograph by George Lee.)

The influx of thousands of sandhill cranes and dozens of endangered whooping cranes has been a major draw for visitors to the refuge. In 2012, the refuge organized its first annual Festival of the Cranes, a one-day event that drew hundreds of visitors. The festival has grown dramatically since and is now a four-day celebration with events on the refuge and in Decatur, drawing over 5,000 visitors. The festival and associated publicity, like this billboard on nearby Interstate Highway 65, have brought added visibility to the refuge and to whooping cranes. (Photograph by George Lee.)

Lizzie Condon (left) and Hillary Thompson from the International Crane Foundation (ICF) release a juvenile whooping crane in the field by the observation building in 2016. This crane delayed its departure from Wisconsin on its annual migration until much later than prudent. ICF personnel trapped it and transported it via van to Wheeler, where it was released and spent the winter. (Photograph by the author.)

Bald eagles have made a comeback nationwide, and Wheeler is no exception. There are at least two active eagle nests on the refuge. In 2005, maintenance worker Jerry Merchant found a young bald eagle that could not fly. Thinking the eagle was injured, refuge personnel captured the bird, and it was transported to the Alabama Wildlife Center at Oak Mountain State Park. Center personnel could find nothing wrong with the bird and suggested refuge personnel search for a nest near the capture site. A nest was discovered nearby—the first documented bald eagle nest on the refuge since 1947. (Photograph by George Lee.)

Thousands of acres of water on the refuge are perfect habitat for osprey. These birds of prey, sometimes called fish hawks, feed almost exclusively on fish. In the 1950s and 1960s, populations decreased dramatically, in part due to the toxic effects of insecticides such as DDT. Possibly because of the banning of DDT in many countries in the early 1970s, together with reduced persecution, osprey populations have made a significant recovery. They are commonly spotted on the refuge, and there are multiple active nests. (Photograph by George Lee.)

In 1964, three otters were trapped on Okefenokee National Wildlife Refuge and flown to the Birmingham airport, where they were picked up by refuge employees. They were transported to Wheeler and released on Beaverdam Creek. The annual report for that year states that they were "probably the first otters to see this locality for over a century," but there were unconfirmed reports of otters prior to those releases. Otter numbers seem to be slowly increasing, and they can now sometimes be seen from the observation building. This recent photograph was taken in Dinsmore Slough. (Photograph by George Lee.)

The author tracks a wayward whooping crane in Tennessee. Many whooping cranes have transmitters on their legs, allowing researchers to track them daily. This whooping crane shortstopped its usual spring migration path from Alabama to Wisconsin and stopped in Tennessee. Using telemetry equipment, volunteers located the crane to determine if it was incapacitated. It was healthy but later became entangled in a wire fence and died. (Photograph by George Lee.)

Three

RECREATION

These birdwatchers gather at White Springs Dike in 1971. Birdwatching is one of the most popular recreational activities on the refuge, and thousands of birdwatchers visit annually. As the number of sandhill cranes has steadily increased in the past two decades along with the increasing numbers of endangered whooping cranes, more and more visitors are showing up to enjoy the spectacle.

Recreational boaters enjoy the Wheeler Lake impoundment created by the construction of Wheeler Dam by the Tennessee Valley Authority. The lake opened up new recreational opportunities to the area, and boaters took to the water as soon as the lake filled. Large crowds like these in 1940 became common.

State game warden Loyd Hays checks out the catch of a local angler on Flint Creek. Hays was later shot and stabbed to death near Flint Creek in Morgan County. He was investigating a vehicle parked near the creek when he encountered a male and female walking nearby. As he approached the pair, the female ran behind Hays for protection and informed him that the male had just raped her. The man then produced a handgun, shooting Hays twice and stabbing him to death.

Two hunters and their dogs are working a field during a quail hunt. Local hunters were attracted to five public hunts (waterfowl, night raccoon, squirrel, quail, and rabbit) held annually to control small game and game bird populations.

A commercial fisherman loads his boat with equipment in preparation for paying out trotlines. Commercial fishing was one of the main economic uses on the refuge and was a lucrative activity in the early decades of the refuge. Tons of fish were caught annually in the Tennessee River and tributaries and were shipped to nearby cities. Individual fishermen were known to catch up to 500 pounds of fish in a day.

Lucky hunters during the 1965 waterfowl hunt display their day's harvest. Harvesting of all waterfowl in-season was allowed during the hunt. Waterfowl hunting was first allowed in 1963 and continued through 1969, primarily due to a high goose population.

Hunters are lined up to get hunting permits at refuge headquarters in 1954. Refuge employees and volunteers issued over 6,000 permits annually.

These duck hunters show off their harvest during the 1968 waterfowl hunt. The construction of Wheeler Lake by the Tennessee Valley Authority created thousands of acres of new habitat for ducks and geese, and hunters followed. The refuge grew to be one of the best waterfowl hunting areas in the region.

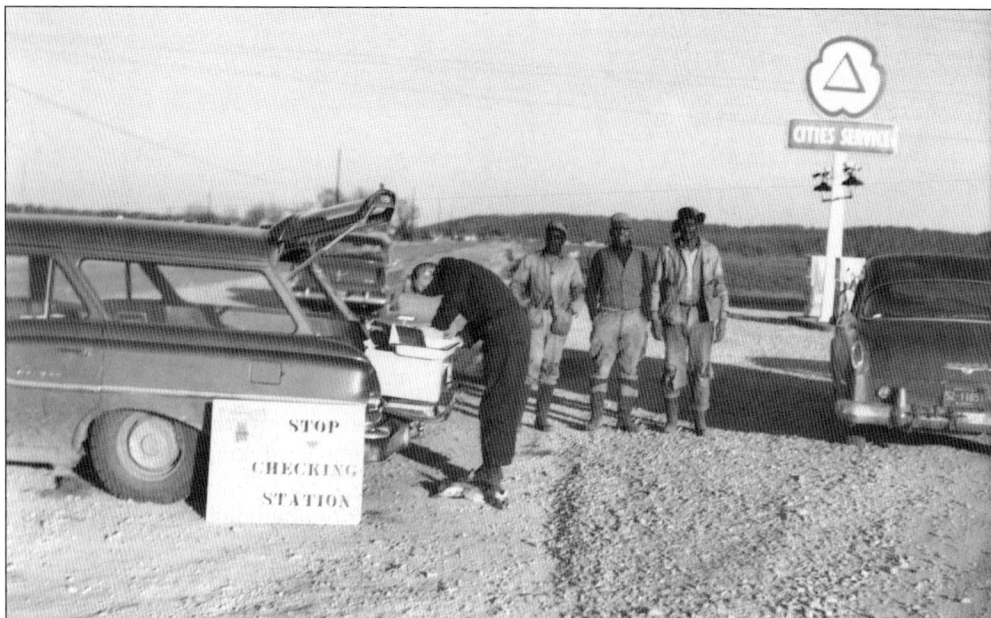

State biologist Ed Hill checks rabbits killed during a refuge hunt in 1963. As the hunters wait, Hill records kill numbers and checks for parasites, pregnancy, and disease. The collected data was used to manage rabbit populations and help set daily bag limits.

Five local youth hunters show off their daily Canada goose limits in 1963. Waterfowl hunting was prohibited on the refuge from 1938 to 1963. These hunters were enjoying the first waterfowl season in many years. In 1969, waterfowl hunting was again prohibited, and remains so today.

A biology class from Florence State Teachers College (now the University of North Alabama) enjoys a visit to the refuge in 1949. Biology classes from local high schools and universities have long utilized the refuge for classes, studies, and field trips. Students from Calhoun Community College, Oakwood University, Alabama A&M University, and local school systems including Decatur, Huntsville, and Hartselle often utilize the refuge for field trips and class projects.

The Huntsville Garden Club is observing waterfowl from White Springs Dike in 1962. Civic, social, and scout groups were frequent users of the refuge. White Springs Dike has long been a favorite area for observing waterfowl. The Tennessee River flows on one side of the dike, and the other side has vast, managed wetland areas that attract huge flocks of ducks. It is on the main road between Decatur and Huntsville and is a short drive from either city.

Local paddlers have always enjoyed the scenic creeks and backwaters on the refuge. Kayaking and canoeing have increased significantly over the years. Limestone Bay, Blackwell Swamp, and Flint Creek are the three most popular paddling spots. The Flint Creek Canoe Trail now offers a 20-mile paddling route through the refuge along Flint Creek to the Tennessee River. Parts of the trail are closed seasonally to avoid disturbance to waterfowl.

In the 1960s, field trials were held on the refuge. Six field trials—three with coon dogs and three with retrievers—were held in 1966. These hunters are preparing for a night coon hunt.

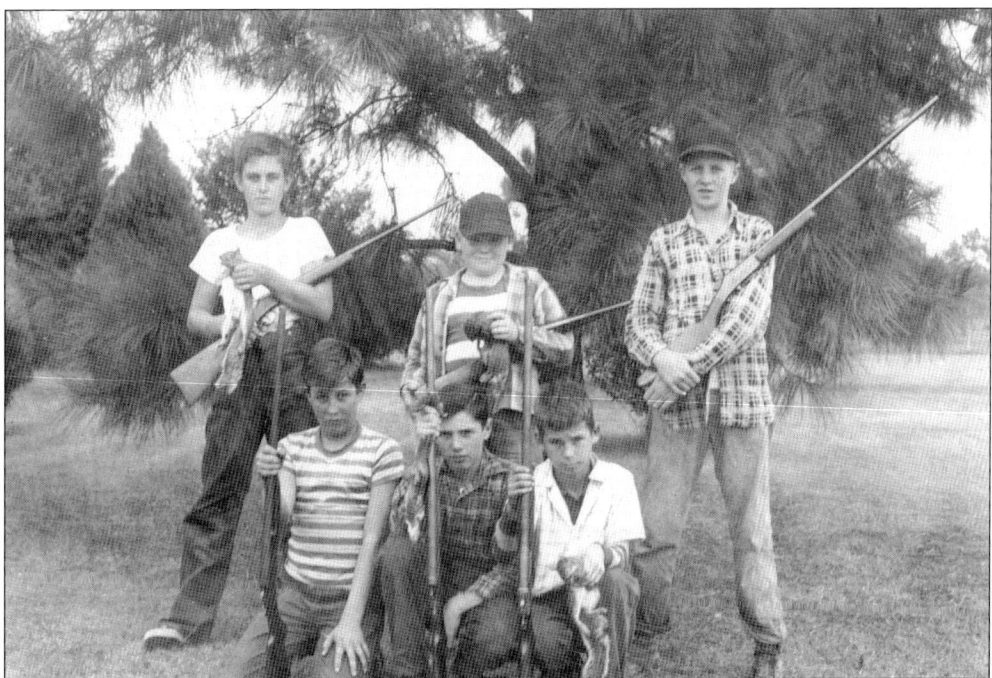

These Junior Squirrel Hunters sponsored by the Birmingham Sportsmen's Association were participants in a refuge squirrel hunt in 1949. Small game hunting, including squirrel hunting, remains a popular refuge activity.

A group of avid birdwatchers is participating in the National Audubon Society Christmas Bird Count in 1969. This annual event still takes place every December on the refuge and is a popular activity. Volunteers and refuge staff have conducted 60 Christmas Bird Counts since 1940. During the peak of the winter migration from December to February, thousands of ducks, geese, cranes, egrets, and herons crowd the wetlands and the Tennessee River, providing optimal viewing opportunities for birdwatchers. Large flocks congregate to feed during the daylight hours and rise up at dusk and head for roosting areas in the shallow backwaters and sloughs to spend the night.

Fishing has always been a popular activity on the refuge. This photograph was taken in 1940 and shows an angler holding a lake sturgeon taken from the Tennessee River. Native to the Tennessee Valley, lake sturgeon were extirpated from the region during the 1960s, partly due to overharvesting—the females were prized for their eggs, also known as caviar. This overfishing, destruction of habitat, and construction of dams decimated the population. In the early 2000s, the Tennessee Valley Authority, the US Fish & Wildlife Service, the Tennessee Wildlife Resources Agency, the Tennessee Aquarium, Tennessee Tech, and other partners began to restock the river, and to date, 170,000 juvenile lake sturgeon have been released in the Tennessee River.

A proud fisherman holds his catch in front of the refuge sign around 1940. Recreational fishing is still one of the most popular refuge activities, attracting thousands of visitors annually.

Maintenance man G.C. Bishop (third from left) is issuing crow shooting permits to a father and son. Refuge manager Tom Atkeson, on the right, looks on. Crow hunting was permitted in 1962 due to a large roost in the Point Mallard area that was so thick, the birds' droppings were killing vegetation.

Anglers crowd the bank of Flint Creek along the Highway 67 embankment immediately west of refuge headquarters during the spring crappie run. A fishing platform now stands east of this area at the Flint Creek Day Use Area and is one of the most utilized fishing areas on the refuge.

This 200-pound, 11-point buck was the largest deer taken on the 1979 deer hunt. By 1979, waterfowl hunting had long been prohibited on the refuge, but deer, small game, and quail and fox hunting were permitted. Deer hunting was allowed with bow or flintlock only.

Bowfishing was permitted on the refuge for the first time in 1968. This man is fishing in the shallow waters in Flint Creek.

The 1969 Wheeler National Wildlife Refuge annual narrative report labels this image "Lady Nimrod." She was one of many hunters to participate in the waterfowl hunt that year. This photograph was taken near the Decatur gate in the White Springs area. The refuge issued hunting permits from a surplus bus set up in the area. Waterfowl hunting is no longer permitted on the refuge. Visible through the trees in the background is a TVA dynamite storage building.

Boy Scout camporees were held on the refuge every year and are still a permitted activity. Youth activities are an important part of the refuge experience. The refuge offers an annual fishing rodeo, summer day camps, and youth-centered educational programs.

An annual Easter egg hunt sponsored by local churches was held on the refuge lawn in the 1960s and 1970s and became a local tradition for a while. These children are egg hunting in front of the original refuge headquarters building.

Some not-so-legal recreational activity sometimes occurred on the refuge. This illicit whiskey still was discovered by refuge personnel in the Swan Pond Bottoms in 1965 and was reported to appropriate authorities. The still was destroyed and removed by authorities. The Bottoms areas are remote and not easily accessible and were favorite haunts for moonshiners.

The annual youth fishing rodeo is the most popular youth activity, with over 350 participants in some years. Youth age 15 and under participate and awards are given to the largest fish caught by species. In the photograph below, the happy anglers display their trophies.

Tally-ho fox hunting was conducted on the refuge for many years from at least the early 1970s up until 2007. Hunts were held in the Mooresville area, Swancott area, and near County Line Road. The hunts were held by local hunt clubs and even employed the services of a huntmaster from England.

Day camps sponsored by the local YMCA provided underprivileged children exposure to nature and the refuge. The refuge offers summer day camp for 8-to-13-year-olds annually. Refuge staff and volunteers hold a series of camps every summer that offer environmental education and outdoor activities for dozens of local youth.

Four

PUBLIC USE AND ENVIRONMENTAL EDUCATION

This service airplane was used to sow ryegrass and wild millet in 1950. Refuge personnel in this photograph are loading ryegrass seed onto the airplane. Airborne operations are limited due to cost but are still used when necessary. At one point, waterfowl census surveys were done from the air, but they are now done from the ground. Most airborne operations today are done by helicopter.

In the picture above, recreationist R.G. Bisbee (left) and biological technician E.N. Waldrep complete the stone base of the large refuge sign erected in 1970 at the intersection of US Highway 31 and State Highway 20. The US Fish & Wildlife Service adopted standardized sign requirements, and all entrance signs were replaced to meet the new standard. All signs were dark brown with the service goose logo, and all lettering painted yellow. The sign in the picture below was located on Highway 67.

Interstate 65 was constructed in the early 1970s and cut through the heart of the refuge. The construction through the refuge was controversial. The planned route took the highway through a part of the refuge that held large numbers of waterfowl. The refuge advocated routing it through the Triana area but was overruled. As a compromise, an agreement was reached that the roadbed would be elevated over the entire span of the refuge and no construction work would occur during waterfowl season to lessen disturbance to waterfowl. This photograph of the roadway pilings that transect the White Springs Dike area was taken in 1971.

This is an aerial view of the interstate where it crosses the Tennessee River in the refuge. This photograph was taken in 1972 from the White Springs Dike area and shows the northbound lanes completed and the southbound lanes under construction. Construction of the Wheeler NWR portion was completed in late 1973.

Maintenance man Gordon Bishop bulldozes a right-of-way for a new road between Blackwell Swamp and Buckeye Slough. This road opened up a portion of the refuge that had previously been reachable only on foot. Illegal hunting was an issue at the time, and Tom Atkeson, the refuge manager, wanted roads to allow for patrolling and monitoring of hunting. Road-building was a near-constant activity on the refuge and continued for years. This road was built in 1963, just 25 years after the refuge was established.

This bulldozer is opening the road to the Pecan Orchard field in 1950. Early maps of the refuge show a multitude of fields marked as "Pecan Orchard," so it is unknown which one this is, but it is typical of the ongoing efforts to open access to remote parts of the refuge.

A refuge truck drives a road on Flint Creek Island in 1950. This is probably what is now called Airport Road, so named because the TVA once had a grass landing strip in the area. The refuge had a limited number of vehicles and trucks like this that served a multitude of roles.

Tupelo logs are being loaded onto a truck at Beaverdam Creek near Highway 20 (now Interstate 565). This photograph was taken in 1966 during an effort to open up the densely wooded area.

Heavy machinery is used to remove logs from a refuge tupelo sale in 1968. Timber sales were held from 1963 to 1994, though none have occurred since.

Forester Richard Bays measures a log harvested from a road right-of-way in the Swancott area in 1963. In the 1960s, late 1980s, and early 1990s, the southern pine beetle adversely impacted forests on the refuge, and yearly salvage sales of timber occurred. The negative effects of the beetle were somewhat offset by the utilization of funds for land acquisition. Funds from receipts of a southern pine beetle sale on the refuge in the early 1990s were used in a timber-for-land-exchange agreement that secured 1,060 acres within the boundaries of Key Cave NWR.

This is a good example of "beating swords into plowshares." The original Ginhouse Branch wooden bridge was rebuilt in 1969 using steel I-beams procured from military excess and heavy steel grating that had once been the catwalk of a Saturn rocket test tower. NASA's Marshall Space Flight Center is located on Redstone Arsenal, which adjoins the refuge.

The last of the mule farmers was local farmer Tom Bibb, a World War I veteran. When farming began on the refuge, the majority of the farmers used machinery drawn by horses and mules. By 1963, when this photograph was taken, farmers had adopted tractors and other mechanized equipment, and Bibb was the only one still using a mule team for his limited farming operation on the refuge.

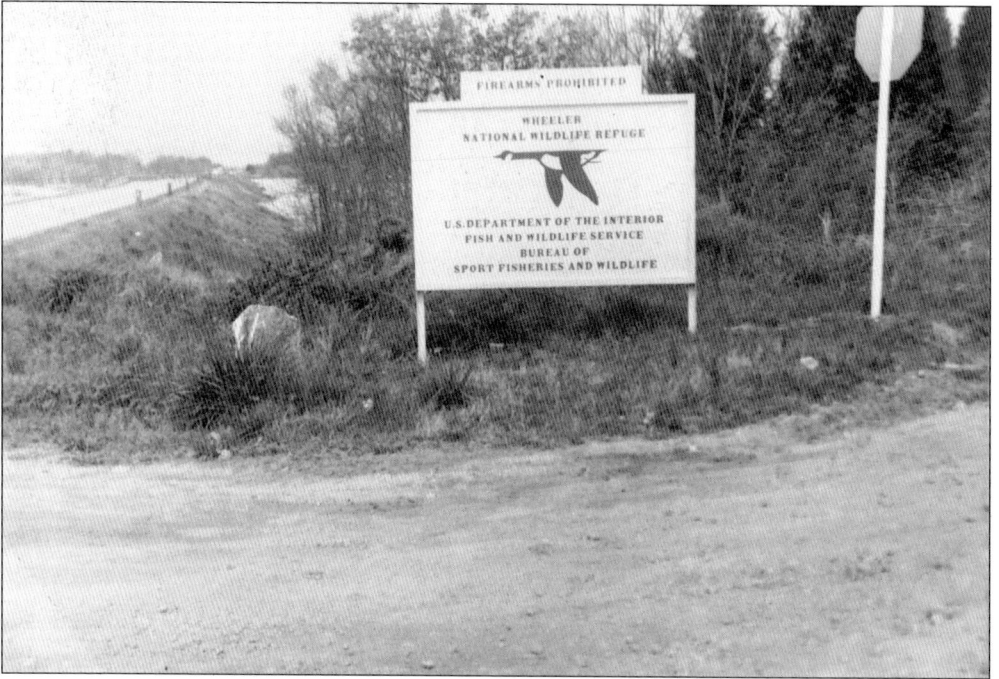

Three new refuge signs were erected in 1961. This sign was on the east side of Flint Creek at the intersection of State Highway 67 and Hickory Hills Road. Hickory Hills Road is on the right, and Highway 67 is in the background at left.

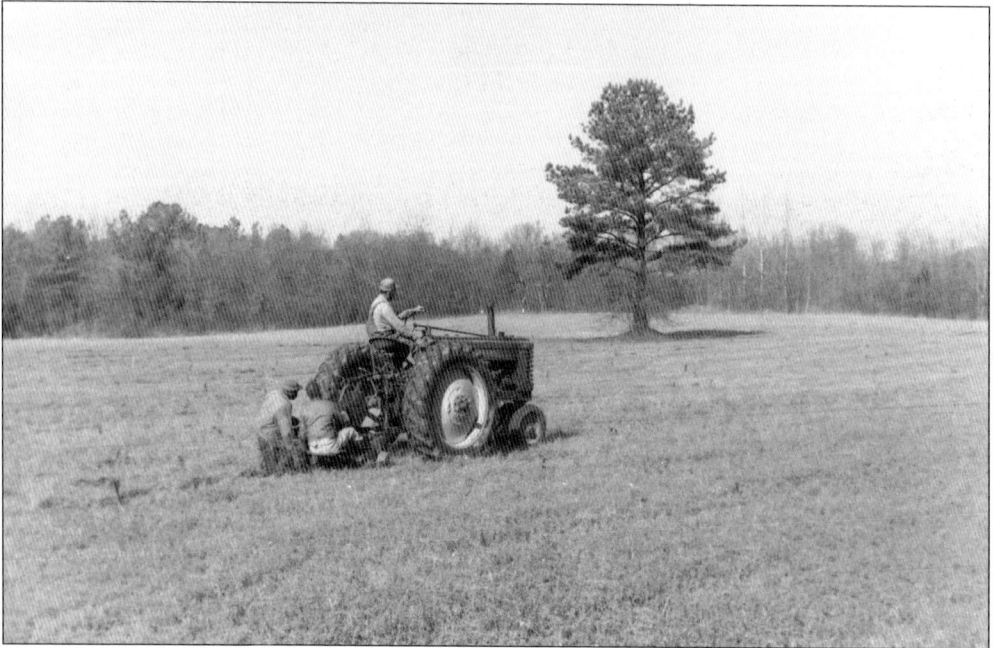

Tree planting was an ongoing endeavor. In the first few years after the refuge was established, as many as 600,000 seedlings were planted annually. Planting continued for decades, and 200,000 pine seedlings were scheduled to be planted as late as the 1963–1964 season. By this time, tractors had replaced hand planting where feasible. This contract crew is planting seedlings in 1964.

The observation building was completed in 1973. This two-story heated building overlooks a large display pool and has one-way glass. Ducks, geese, whooping cranes, and sandhill cranes often congregate by the thousands in the pool during the winter migration. This is arguably the best place in Alabama to view large numbers of waterfowl up close.

Earthmoving equipment and trucks are moving dirt as work on the display pool dike nears completion in 1972. The observation building overlooks this pool. A photography blind on one end of the pool is often used by photographers to capture images of some of the thousands of ducks, cranes, and geese that congregate here in the winter.

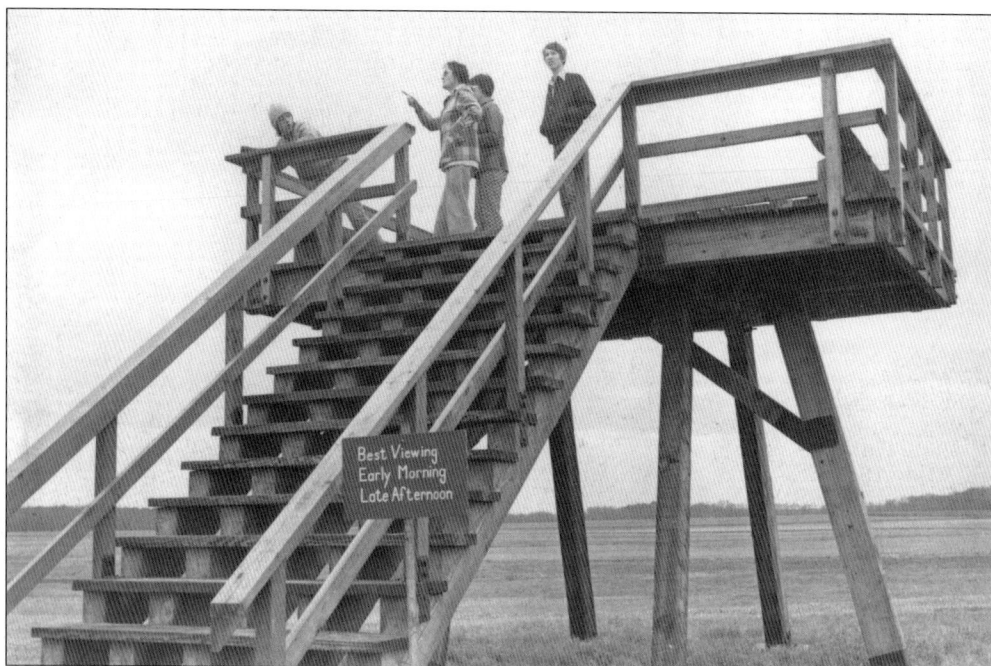

The Beaverdam Peninsula viewing platform was constructed in 1975 on the north side of the Tennessee River near Mooresville. This elevated platform overlooks open fields and is a popular spot for viewing sandhill cranes, snow geese, and whooping cranes in winter. These birdwatchers are looking for waterfowl in the surrounding fields.

The Atkeson Cypress Trail was constructed behind the visitor center in the late 1970s through a cypress swamp and terminated at this raised observation platform. The platform was later removed, and the trail was extended past the cypress swamp into a longer loop through a wetland area.

In 1941, over 4,000 acres of the refuge were incorporated inside the US Army's Redstone Arsenal boundary for security reasons. After the end of World War II, the arsenal became the focus for the development of the country's space program. Wernher Von Braun and his team of German scientists led the space program on the arsenal. Much of the early development of the first rockets was conducted here. Currently, about 1,500 acres of the 4,085 acres are administered by NASA's George C. Marshall Space Flight Center. This photograph shows a refuge wetland with a space shuttle engine test stand in the background.

Refuge employees load seed onto a helicopter for mudflat planting in 1976. The helicopter was provided by the TVA and was used to sow 110 acres of oats to supplement goose forage. Many different techniques have been used to supplement forage for migrating birds. In recent years, cooperative agreements with farmers have been the norm. Local farmers plant feed crops and leave 20 percent in the fields for waterfowl forage. The arrangement benefits the farmer, the refuge, and the birds.

The Youth Conservation Corps (YCC) has been an important resource for the refuge over the years, providing valuable manpower for refuge maintenance and projects. This YCC crew in 1971 is clearing willows from the White Springs Dike area.

Service fish biologists John Forester and Jim Stewart remove grass carp from Thomas Spring in Watercress Darter NWR in 1981. Grass carp negatively impact habitat for the endangered watercress darter, a small fish. Located near Bessemer, Alabama, the refuge was established in 1980 to provide protection for the darter. The refuge is 24 acres of ponds and mixed pine-hardwood forest and contains Thomas Spring, one of six known habitat locations for the endangered darter. A second pond was constructed on the refuge in 1983 to provide additional darter habitat. Watercress Darter NWR is part of the Wheeler NWR complex and is managed by Wheeler.

Southern pine beetles spread through many southern pine forests, and Wheeler NWR is no exception. Beetle-infested trees were removed in an effort to stop their spread. This removal operation took place in 1988. Removal of the downed trees in sensitive areas was restricted to horse and mule operations to reduce damage to the remaining trees.

Refuge employee Anita Bowman holds an injured juvenile tundra swan while the cameraman and soundman for CBS *Sunday Morning*, hosted by Charles Kuralt, record the scene.

A celebration of General Wheeler's 168th birthday was held at the Wildlife Interpretive Center in 1982 and included the firing of muskets and reenactors in Civil War uniforms. Maj. Gen. Joseph "Fighting Joe" Wheeler was a Confederate general, served in the US Congress afterward, and as a general in Cuba with Theodore Roosevelt and his Rough Riders in the Spanish American War in 1898. He played a significant role in the capture of San Juan Hill. Pond Spring, Joe Wheeler's home, is located nearby in north Alabama. Many local landmarks, including Wheeler Dam and Joe Wheeler State Park, are named in his honor. The refuge is named after Wheeler Lake, on which much of it sits, so it is indirectly named after him.

In 1987, the *Delta Queen* riverboat made a stop in Decatur as it made its way along the Tennessee River, and outdoor recreation planner Harvie Fowler gave an offsite educational program to the passengers.

Ranger Daphne Moland holds Hawkeye, a red-tailed hawk. Hawkeye was a well-known and popular member of the refuge team. She was unable to fly due to injuries suffered in a collision with a vehicle and could not be released back into the wild. She delighted visitors for 13 years until her passing in 2013.

At top, Tom Atkeson (left) is pictured at his retirement ceremony with US Fish & Wildlife Service director Frank Dunkle in 1987. Wheeler NWR has been fortunate to have many dedicated leaders over the years. Perhaps the most memorable manager, one whose name is inextricably intertwined with the refuge, was Atkeson. He was one of Wheeler's first employees, starting as a junior biologist in May 1939, only months after the refuge was established. One of his first assignments was to map the entire refuge, an effort that was to prove fortuitous. Atkeson joined the Army during World War II and was injured by an antitank mine explosion while training at Fort Hood, Texas. The accident blinded him and severed both hands. Remarkably, after recuperating from his injuries, he returned to work at Wheeler in 1946, where his 1939 mapping assignment turned out to be a godsend. Using his recollection of the mapping effort as a starting point, he memorized the refuge's roads and trails. His knowledge of the refuge became so intimate that he could tell his location by the sounds of the roadways and by the number of turns his (driven) vehicle took. He rose steadily through the ranks and became refuge manager in 1962 and held that position until 1987. Charles Kuralt filmed a television tribute to him for CBS Sunday Morning in 1984. Tom Atkeson is pictured at bottom in his Army uniform.

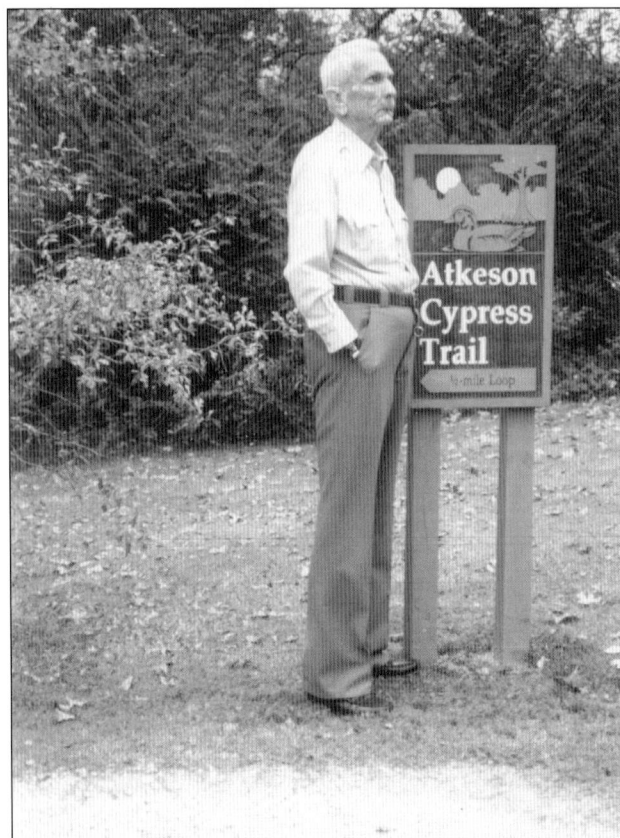

Refuge personnel gather at the opening of a nature trail at the headquarters area in 1966. From left to right are refuge manager Tom Atkeson, biological technician E.N. Waldrep, soil conservationist J.L. Derden, administrative assistant H.L. Fowler, and maintenance men G.C. Bishop, J.H. Blackwood, and T.P. Sandlin.

In 1989, former refuge manager Tom Atkeson was recognized by naming the Atkeson Cypress Trail in his honor. Atkeson retired in 1987 after 48 years of service with the US Fish & Wildlife Service, including 25 years as Wheeler's refuge manager. The trail is located next to the visitor center and is a popular nature walk for visitors.

STANDARD FORM NO. 64

Office Memorandum · UNITED STATES GOVERNMENT

TO : Mr. Thos. Z. Atkeson,
    Thru' Regional Director, Atlanta, Ga.

DATE: June 17, 1949

FROM Acting Chief, Division of Information,
    Washington, D. C.

SUBJECT: Conservation in Action booklet: Wheeler,
    a National Wildlife Refuge.

Your publication on the Wheeler Refuge has just been received from the Printing Office and I want to congratulate you on an outstanding addition to the Conservation in Action series. Your manuscript was not only full of interesting and useful information, but was extremely well written. I thoroughly enjoyed reading every bit of it and I am sure that it will be well received. Mr. Hines, also, did an excellent job on the illustrations, making this one of our most attractive publications.

As perhaps you know, you are entitled, as author, to 75 copies. We will be glad to forward these to you or to mail out single copies to a list supplied by you.

We are looking forward to other numbers from you, and hope that we will not have to wait too long.

Rachel L. Carson

Rachel L. Carson

Attachment

Rachel Carson sent this memorandum to Tom Atkeson in 1949. Carson worked for the US Fish & Wildlife Service and later became famous for writing *Silent Spring* in 1962. In her book, Carson addressed the adverse effects of DDT and other chemical pesticides on birds and other wildlife, citing among others a 1952 incident on Flint Creek. Carson is often credited with the birth of the modern environmental movement.

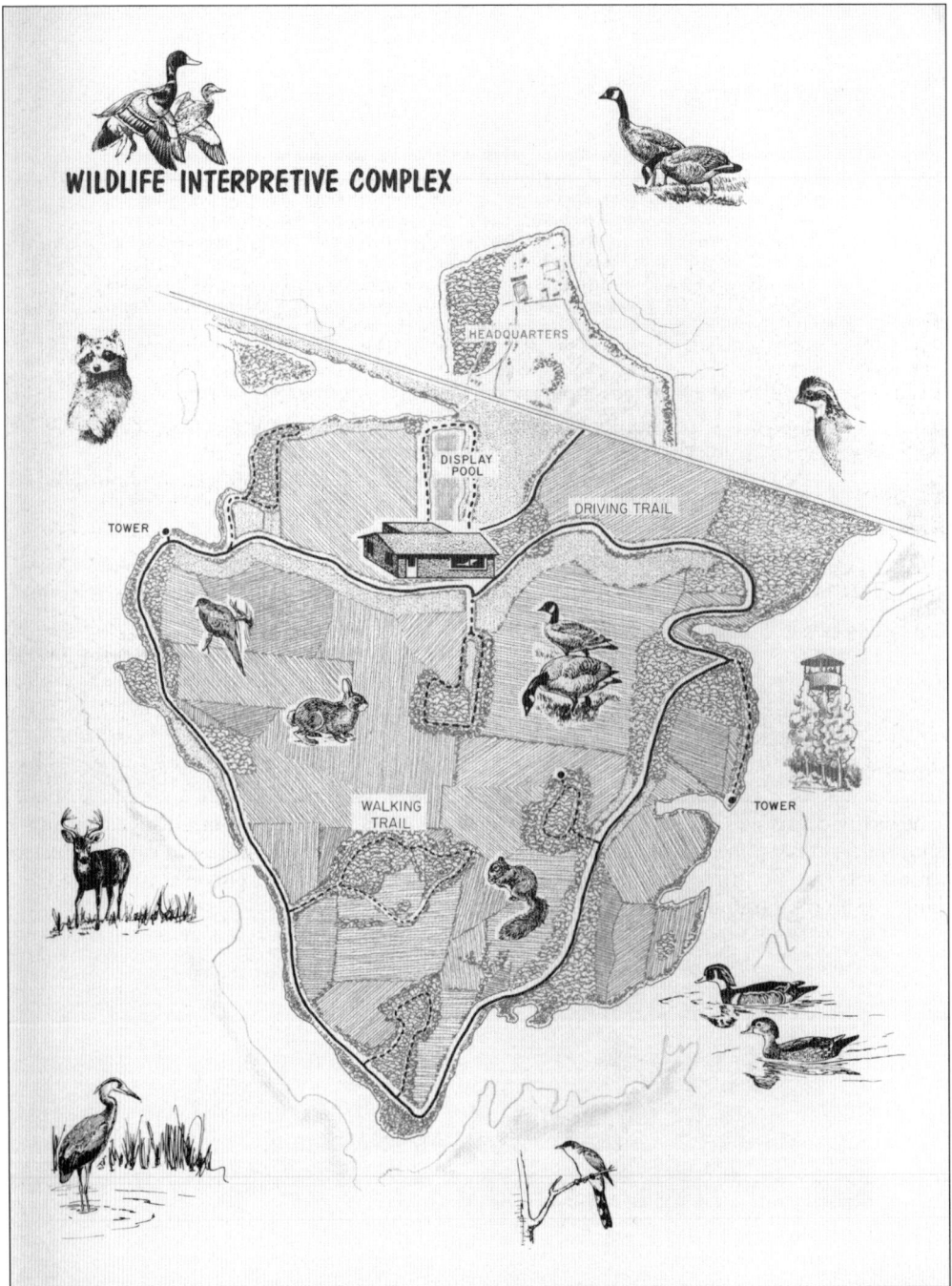

WILDLIFE INTERPRETIVE COMPLEX

HEADQUARTERS

DISPLAY POOL

DRIVING TRAIL

TOWER

WALKING TRAIL

TOWER

This is the map with the original plans for the proposed Wildlife Interpretive Center (currently called the visitor center). The plan called for a driving trail and observation towers, but those were never developed. The visitor center complex today includes an observation building and the Tom Atkeson Cypress Trail and boardwalk.

The visitor center was constructed and opened to the public in 1980. The original cedar siding is shown here. The siding was replaced during a later renovation. The center significantly increased visitor services. Prior to its construction, limited visitor facilities were provided in management buildings in the headquarters complex.

In November 1982, the Wildlife Interpretive Center (the current visitor center) was rededicated and named for Lawrence S. Givens. Givens was refuge manager from 1945 to 1949. Jane Givens (second from right) and family members are shown near the new sign at the dedication ceremony.

The Lawrence S. Givens Visitor Center main lobby is shown with displays after a renovation in 1988. The renovation included the upgrade of the dome displays in the center of the room, which exhibit reptiles and amphibians, and the mounted displays of a kestrel, raccoon, and others along the left wall.

This waterfowl display in the visitor center was also renovated to add additional species. These displays are still used today.

Two open waterfowl dioramas were added during the renovation. The dioramas display waterfowl and wildlife endemic to the refuge.

This is evidence of a not-so-desirable public use of the refuge. Assistant refuge manager Randy Cook stands by a confiscated marijuana plant in 1985. Two individuals were arrested for growing and possessing marijuana on the refuge.

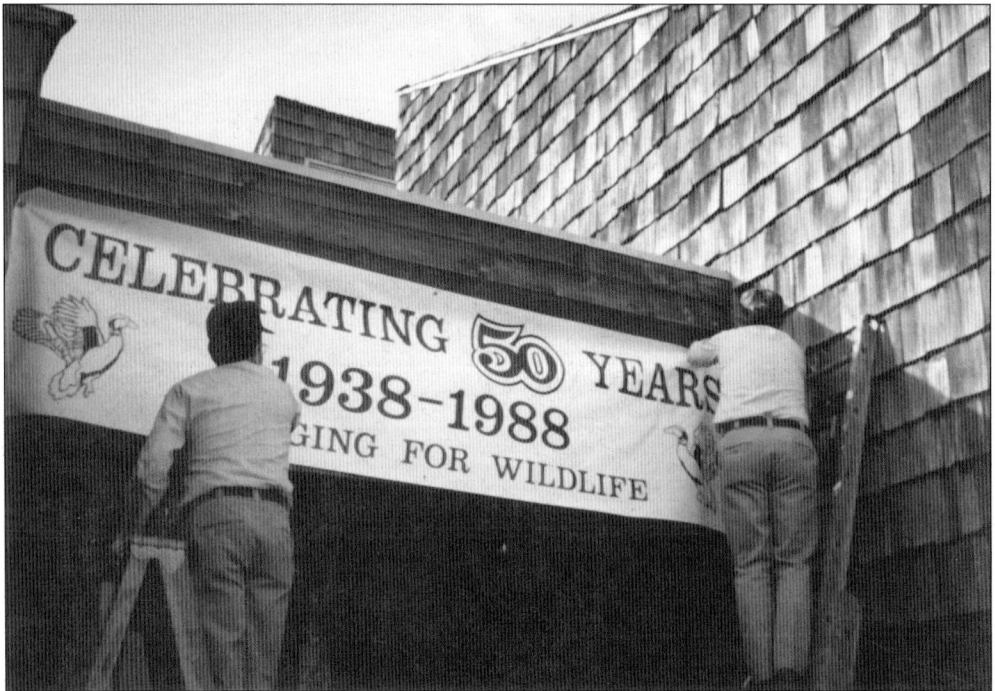

Harvie Fowler and Tuck Stone hang a 50th anniversary banner on the front of the visitor center in 1988. The event was sponsored by the Tennessee Valley Audubon Society with support from local organizations and companies. The celebration events included a bird-watching tour, birdhouse-building classes, and archery and muzzleloader firearms exhibits. Guest speakers included former refuge manager Tom Atkeson and USF&WS regional director Jim Pulliam.

The Walt Disney Pictures production of the movie *Tom and Huck* was filmed at various locations on the refuge in 1995. From left to right, this photograph shows the three main characters, Huck (Brad Renfro), Injun Joe (Eric Schweig), and Tom (Jonathan Taylor Thomas), on the set at Mussel Camp.

This tavern and dock set for the movie was constructed at Mussel Camp. The building on the left was bought by a local citizen and moved off of the refuge after filming ended. It now sits just off the refuge boundary on Bluff City Road.

Excavation of a Paleolithic site called Beartail Rock Shelter was undertaken in 1995. This site contains archeological artifacts that reveal human occupation of the site spanning 10,000 years. A University of Alabama archeological excavation crew discovered stone implements and tools and reserved organic material including bones and shells. This is a rare discovery; only a half-dozen other such sites have been discovered in the eastern United States.

Native American artifacts are sometimes found in the backwaters and along the banks of the Tennessee River. The Tennessee River valley was a hunting ground for the Creek, Chickasaw, Choctaw, and Cherokee tribes. Artifact collecting is illegal on the refuge. These artifacts were confiscated from two men apprehended while illegally collecting on the refuge. Together, they had 91 high-quality artifacts.

This is a c. 1950 aerial view of the headquarters complex. It includes two residences, shop and service buildings, a garage, and associated buildings. Twelve years after the establishment of the refuge, the land still looks barren compared to today.

The refuge entrance at the headquarters complex is pictured here in 1955. Tree planting efforts are just starting to show results. This entrance along State Highway 67 now houses the modern headquarters building, a bunkhouse residence, and associated maintenance buildings. The area is now surrounded by a shady grove of mature hardwoods and pines.

The Junior Duck Stamp contest judges pictured here are, from left to right, Dwight Cooley, unidentified, Sam Hamilton, Barnett Lawley, and Vic Daily. This annual contest is conducted throughout Alabama, and judging takes place at Wheeler NWR. Winners are selected by age group from all art submitted, and an overall state winner is also selected. The overall winner is then judged at the national level, and a final national winner is selected. Cooley was deputy refuge manager from 1997 to 2001 and refuge manager from 2001 to 2016.

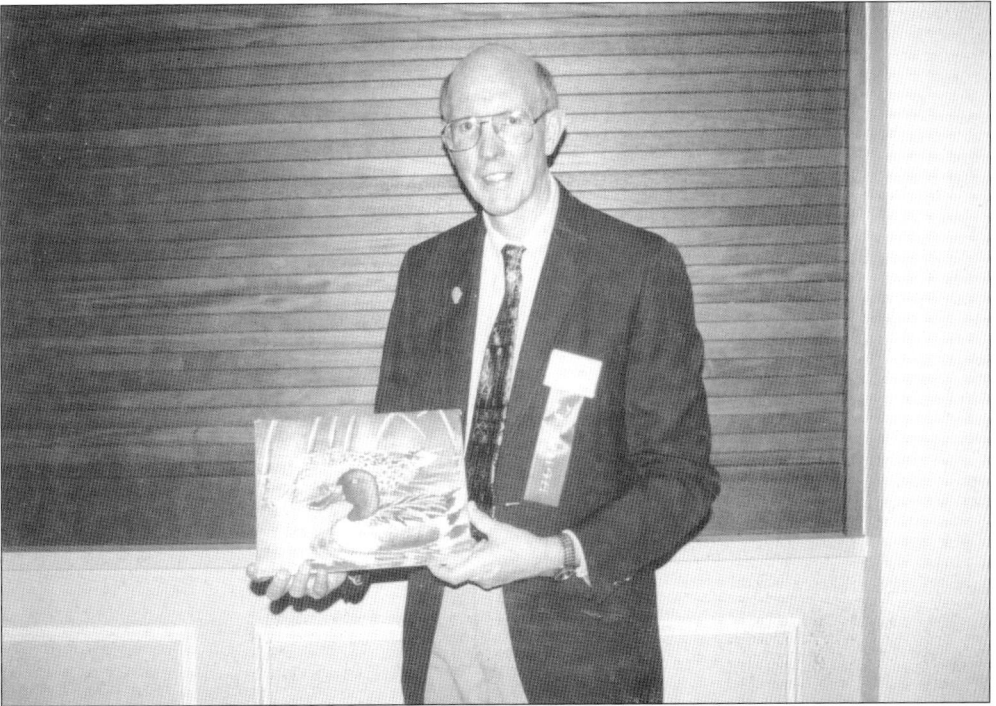

Above, Troy Wyers of Ducks Unlimited holds the winner of the 2000 Alabama Junior Duck Stamp contest. The artistic rendering of a pair of northern shovelers was chosen from a field of entries from all over the state and went on to compete in the national contest.

WHEELER

National Wildlife Refuge
Limestone, Morgan, and Madison Counties   **Alabama**

WHEELER

This is the cover of a refuge brochure produced in 1969 to provide public information on the refuge and future plans for recreation and development. The brochure included maps and plans for future trails and observation areas and a historical accounting of the refuge and its development to date.

Canada geese are captured using a rocket net. The net was remotely triggered to catch large flocks of geese at one time. Geese were tagged and released to track population and movement.

Assistant manager Ernest Jemison demonstrates duck banding to a local youth group. The metal bands on the necklace around his neck were applied around the leg of the duck. Each band has a number and telephone number. When a banded duck is found and the band number reported, biologists can determine the age and migration path of the bird.

Refuge employee Harvie Fowler inspects a display of mounted waterfowl in the old headquarters building. The refuge has an extensive collection of waterfowl mounts on display in the visitor center.

Five

WHEELER NWR COMPLEX

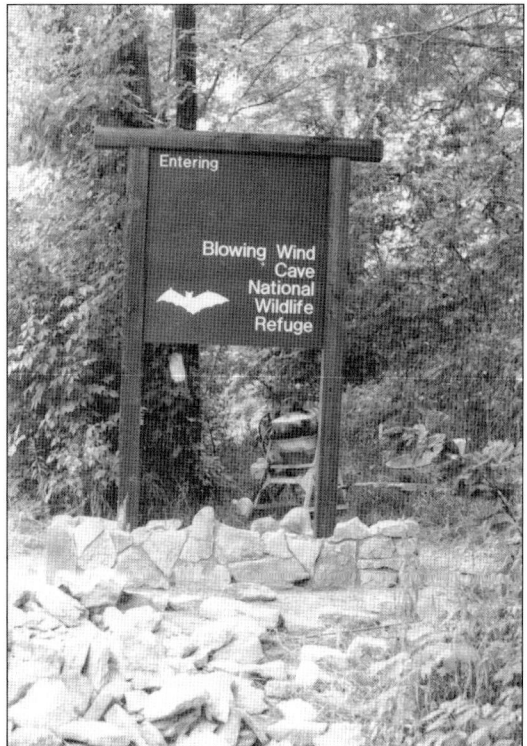

Wheeler's mission to protect wildlife has always been foremost, and over the years, six noncontiguous satellite refuges were established that are managed by the refuge staff. These are Sauta Cave in Jackson County (added in 1978), Fern Cave in Jackson County (1981), and Key Cave in Lauderdale County (1997), all three of which were established to protect specific endangered species, plus Cahaba River near Birmingham (2002), Watercress Darter in Jefferson County (1980), and Mountain Longleaf in Calhoun County (2003), which protect unique ecosystems. Collectively, these refuges make up the Wheeler NWR Complex. This is the entrance sign to Sauta Cave NWR (known as Blowing Wind Cave NWR until 1999).

Members of the Wittenberg University Speleological Society from Springfield, Ohio, visit Fern Cave NWR in 1987, accompanied by Fred Bagley, endangered species office and refuge personnel. Fern Cave is home to the largest wintering colony of gray bats in the United States. Cavefish and cave salamanders also live there. Fern Cave NWR is a satellite refuge located on a 199-acre tract 55 miles west of and managed by Wheeler NWR headquarters. It is a forested hillside underlain by a massive cave with many stalactite and stalagmite-filled rooms. Fern Cave NWR was established in 1981 to provide protection for federally endangered gray and Indiana bats. As many as one million gray bats use the cave as a hibernaculum, as do several hundred Indiana bats. The cave has been closed to all public entry since 2009 due to concerns about the spread of white-nose syndrome, a fungus that has resulted in 95 percent–plus bat mortality in some hibernacula. The federally threatened American Hart's-tongue fern occurs at one of the entrances.

In 1992, the efforts of Steve Pitts, a member of the Huntsville Grotto of the National Speleological Society, resulted in the discovery of Pleistocene-era bone material, including this human mandible, in a remote area of Fern Cave.

Other Pleistocene bones, from the giant armadillo, giant peccary, jaguar, cave bear, and horse, as well as the toe bone of a giant ground sloth and a deer skull and antler, were also found by Steve Pitts. These artifacts are currently at the USF&WS regional office in Atlanta.

One of the picturesque stalactite- and stalagmite-rich caverns in Fern Cave is shown here. The cave has five known entrances, four of which are on the refuge. Access is extremely difficult, and the cave has been described as a vertical and horizontal maze by expert cavers. Horizontal sections of the cave are known to be more than 15 miles long, and vertical drops of 450 feet are found within. The cave is closed to the public.

In 1990, a group of spelunkers from Russia visited the refuge and explored Fern Cave. This picture shows the five Russian spelunkers with six members of the Huntsville Grotto, National Speleological Society, in front of the cave entrance. The caves were popular destinations for foreign cavers, and groups from other European countries came to explore them.

Cave Springs is a major cave system on the refuge. Pictured here is one of the caverns within the system. Cave Springs supports a significant population of endangered gray bats. In 1986, the cave experienced a large bat die-off. Analysis of bat guano collected from the cave revealed high concentrations of DDE and DDT. The chemicals have been found to adversely affect bird and bat populations. The sale and use of both chemicals have since been banned.

Biologists Fred Bagley (left) and Ron Freeman are seen at the upper entrance to Sauta Cave. The cave was used for a variety of activities before being protected as a national wildlife refuge. Refuge employees found a large cast-iron pot used for saltpeter production in the cave. Sauta Cave NWR, also a satellite refuge managed by Wheeler NWR, is located 60 miles east of Wheeler NWR.

Visitors observe bats emerging from Sauta Cave near dusk. Sauta Cave hosts what is believed to be the largest summer gray bat concentration east of the Mississippi River plus much smaller numbers of Indiana bats, both endangered species. The refuge's 264 acres were purchased in 1978 to provide protection for the endangered bats. The cave is a summer roosting site for about 200,000–400,000 gray bats and a winter hibernaculum for both the gray and Indiana bats. There are two entrances on the refuge, but they are closed to the public to limit disturbance. In addition to the rare fauna within the cave, the federally threatened Price's potato-bean occurs on the refuge. The Alabama Natural Heritage Program ranks Sauta Cave as a site of very high significance.

Surprisingly, Sauta Cave is not pristine, as it was used as a saltpeter mine during the Civil War, a nightclub during the 1920s, and a fallout shelter during the 1960s. This picture shows a dance hall on the site in the mid-20th century.

This fish camp was in operation near the cave in the 1960s. Note the "Sauty Cave" spelling on the sign. Locals often called the cave by that name.

This log dwelling on Sauta Cave NWR is believed to date to the 19th century. It was occupied for years by a series of caretakers who lived onsite and whose presence prevented vandalism, cave intrusion, and litter on the refuge. The house is still standing.

A bat is examined by biologists before attaching a radio-telemetry device or transmitter. The tracking was part of a study led by Auburn University to determine the effects of vegetation removal from nearby Lake Guntersville. The transmitter monitored bat movement and foraging patterns. Transmitters were attached to 27 bats during the study.

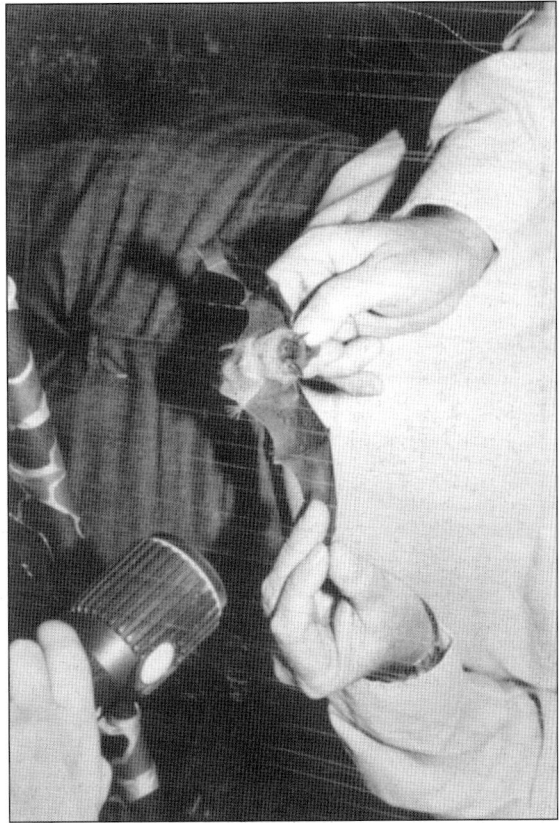

Bats are shown flying toward the entrance to Sauta Cave. On summer evenings, visitors gather on the observation platform in front of the cave to watch the emergence of thousands of bats from the cave mouth. The platform is located just in front of the mouth of the main cave entrance.

A cluster of gray bats hangs from the Sauta Cave ceiling. The population fluctuates based on environmental and natural factors. In 1985, the population was estimated at 485,000 and has remained relatively stable. A recent threat to the cave's bats is the detection of white-nose syndrome in the cave. The impact on Sauta Cave bats is unknown at this time.

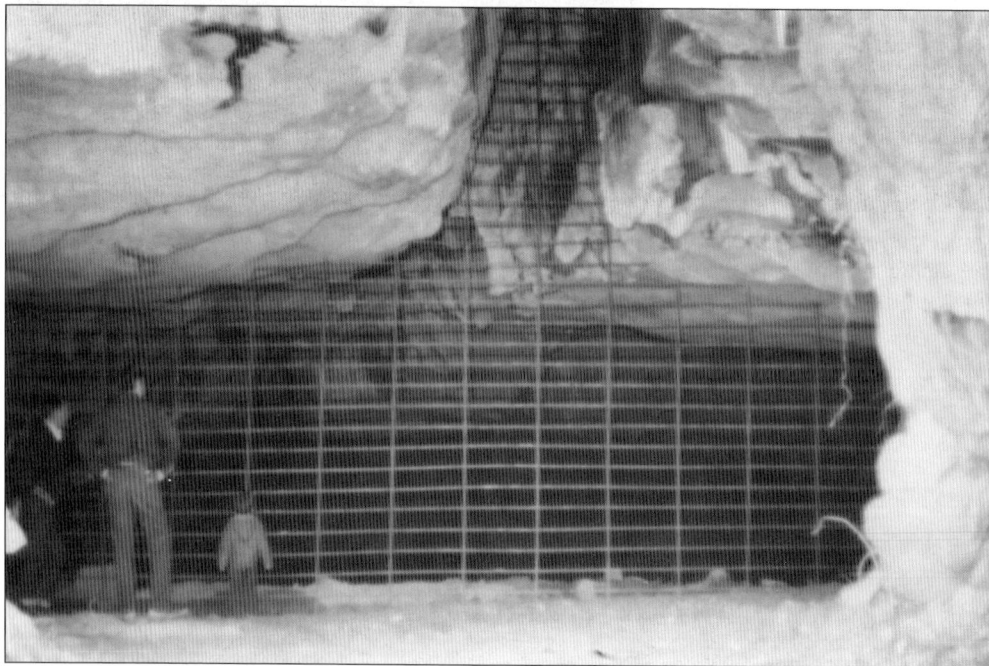

Biologists Fred Bagley and Ron Freeman prepare to enter Sauta Cave for an inspection. This photograph was taken in 1986, shortly before two men damaged the barricade across the cave mouth and breached the entrance. The on-site custodian spotted them and reported their vehicle tag number. They were arrested and convicted. Human disturbance in past years greatly reduced the population of the bat colony. Public entry is prohibited in order to avoid disturbance to endangered bats.

Refuge personnel conduct annual public programs at Sauta Cave where visitors can witness the bats emerging at dusk for their nightly hunt. Here, refuge employee Darrin Speegle, in a hard hat, discusses the cave and bats with a group of Audubon Society and Sierra Club members in 1988.

Key Cave NWR, established January 3, 1997, is a 1,060-acre tract of rolling hills and cropland. The refuge and the area directly north of it is part of the groundwater recharge area for Key Cave, the only known location of the Alabama cavefish. Consequently, the cave is designated as critical habitat for the endangered Alabama cavefish and for the endangered gray bat. The refuge is managed for a variety of habitats including native grasses, hardwoods, grassland or hedgerow habitat, shallow-water areas, and forested land dominated by upland oaks and hickories. Prescribed fire is the principal tool used to manage refuge habitats for grassland bird species at Key Cave.

The 9,016-acre Mountain Longleaf NWR was established May 31, 2003, within the former military training base of Fort McClellan. Approximately one third of the refuge is open to the public, with the remainder undergoing unexploded ordnance removal and other environmental contaminant cleanup. The primary objective of the refuge is to maintain and restore a naturally regenerating mountain longleaf pine ecosystem.

Cahaba River NWR was established September 25, 2002, for the purpose of protecting and managing a unique section of the Cahaba River and land adjacent to it. The 3,690-acre refuge and surrounding areas are designated a Significant Landscape for longleaf pine conservation and management.

Scattered along the Cahaba River is a series of rocky shoals populated by Cahaba lilies. The lilies begin flowering in late April and continue into late June. The prime flowering season typically occurs from early May through the third week of June. Each year, the Cahaba Lily Festival is held on the third Saturday in May in nearby West Blocton, Alabama.

This is the Wheeler Wildlife Refuge Association logo. (Courtesy of Wheeler Wildlife Refuge Association.)

EPILOGUE

The national wildlife refuge system consists of over 550 refuges. The mission of the refuge system is "to administer a national network of lands and waters for the conservation, management, and where appropriate restoration of fish, wildlife, and plant resources and their habitats within the United States for the benefit of the present and future generations of Americans." This is bureaucratese for preserving lands for the benefit of wildlife and nature. Each refuge is unique in its own way. Whether it protects habitat for sea turtles (Archie Carr NWR), key deer (National Key Deer NWR) or caribou (Arctic NWR), there is a reason each refuge exists.

For Wheeler NWR, the primary reason is to provide refuge for waterfowl. But in addition to providing habitat for over 80,000 migratory waterfowl, the refuge boasts 115 species of fish, 74 species of reptiles and amphibians, 47 species of mammals, 300 species of songbirds, and 13 federally endangered or threatened species. It is an island of nature amidst an ever-expanding sea of encroaching development.

As the area's population continues to grow, the value of the refuge will only increase. Wheeler does not charge entrance fees, and almost all events and activities are free of charge. Like all NWRs, Wheeler is almost entirely tax-supported. Although polls show that the public overwhelmingly supports wildlife refuges (as well as almost all of the lands administered by the Department of Interior), federal budgets continue to lag well behind the level needed to maintain a viable refuge system. With the recent and continuing budget cuts to the Department of Interior, our refuges struggle to meet their operational needs and provide services to the public.

These budget shortfalls affect refuges in two ways. Most importantly, infrastructure needs are not met, resulting in facilities deterioration. Trails, boardwalks, fishing piers, visitor centers, and roads fall into disrepair as staff struggle to maintain a safe and acceptable level of use. Second, visitor services are curtailed, and educational programs, public events, and outreach are limited.

To fill this funding gap, refuges rely on the support of volunteers and nonprofit partners. Most refuges, Wheeler included, greatly depend on volunteers to fill staff shortfalls that result from budget cuts. Wheeler has a cadre of over 40 part-time volunteers who staff the visitor center, mow grass, perform maintenance tasks, and assist in biological surveys. Without these volunteers, many functions of the refuge could not be performed.

Wheeler's nonprofit partner is the Wheeler Wildlife Refuge Association (WWRA). This all-volunteer "friends" organization assists in events and programs, raises funds, and manages the on-site gift store. The WWRA is an integral part of the refuge's operation and sustainment. WWRA relies on membership contributions and dues for a portion of its funding, and all WWRA proceeds are used for refuge support.

We encourage you to become involved with Wheeler NWR. Please consider volunteering your time and talents or joining the Wheeler Wildlife Refuge Association. Contact the refuge office and help your refuge continue to be a remarkable place for wildlife and people.

One hundred percent of the author's proceeds from the sale of this book go to the Wheeler Wildlife Refuge Association.

DISCOVER THOUSANDS OF LOCAL HISTORY BOOKS FEATURING MILLIONS OF VINTAGE IMAGES

Arcadia Publishing, the leading local history publisher in the United States, is committed to making history accessible and meaningful through publishing books that celebrate and preserve the heritage of America's people and places.

Find more books like this at
www.arcadiapublishing.com

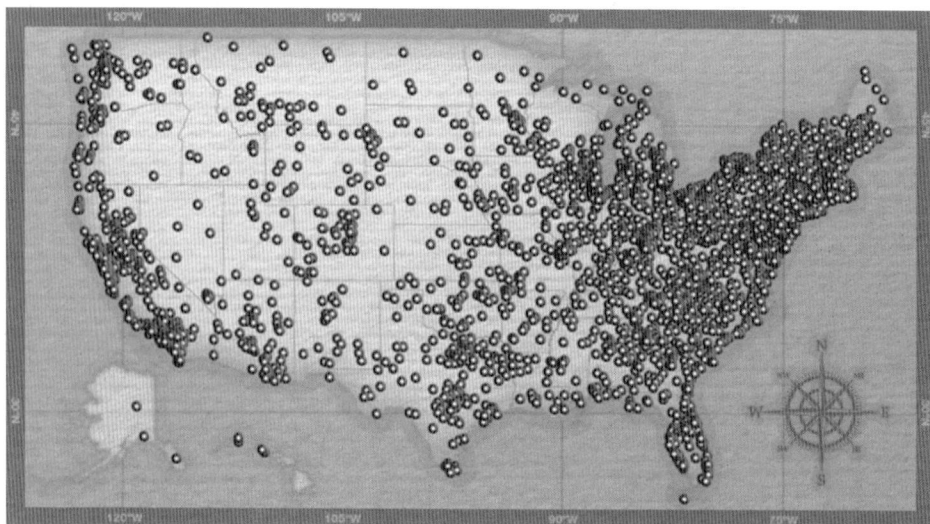

Search for your hometown history, your old stomping grounds, and even your favorite sports team.

Consistent with our mission to preserve history on a local level, this book was printed in South Carolina on American-made paper and manufactured entirely in the United States. Products carrying the accredited Forest Stewardship Council (FSC) label are printed on 100 percent FSC-certified paper.

MADE IN THE USA